Horse Power

By
Frank Lessiter

"I have always considered that the substitution of the internal combustion engine for the horse marked a very gloomy milestone in the progress of mankind..."
—Sir Winston Churchill

**Layout and Design By
Keith Bush**

**Edited By
Roy Reiman**

$11.95
Reiman Publications, Inc., Milwaukee, Wisconsin

This Book's Pedigree
(Our Table of Contents)

Iowa farmer Dick Sparrow and his 40-horse hitch. . . . . See page 10.

When "horsepower" meant real horses and not cars . . . . See page 40.

How draft horses had to change with the times on the farm. . .See page 51.

6 "Giddap!" (Foreward to Our Book) . . .

The book's author explains what sparked this special book on draft horses, and relates a few of his childhood experiences with big horses. He also shares with readers some unmatched memories—both as a boy growing up on a farm where draft horses were used, and in later years when he began gathering material for this book.

10 He Drives the Most Famous Hitch of All—"The 40" . . .

No modern-day book on draft horses would be complete without a chapter or two on Iowa farmer Dick Sparrow, the man who put together the first 40-horse hitch to pound the pavement since the country's big circuses retired their big hitches back in 1904. Sparrow vowed that before he quit driving the "40" he would leave something in writing to help those who might try again to drive such a hitch in the future. He does that in this chapter, explaining his methods and ideas on driving 40 horses.

37 $800 of Groceries Each Week . . .

That's what it took to feed the crew and friends when Dick Sparrow's 40-horse hitch was on the road. Buying better than $12,500 in groceries in strange away-from-home supermarkets during 16 weeks on the road, the two gals who did the cooking had their hands full. "It was like feeding an old-time threshing crew three meals a day," they say.

40 When The Draft Horse Ruled the Land . . .

For hundreds of years, draft horses were the main means of transporting people and goods in the new world. Hay was the "gasoline of the times" as horses helped this country grow.

51 Memories of Real Horsepower . . .

The author remembers the teams of horses used on his own home farm . . . and some of the interesting changes in horsepower that were made by his dad over the years. Ever hear of a horse being allergic to hay? That story is told here, too.

54 "Belgian Is My Breed" . . .

An avid owner of Belgians tells why he's partial to this breed. This is the first of seven chapters on various draft horse breeds. In each chapter, an outstanding owner of a different breed describes his breed preference and details the background of his particular breed. *Each breeder also shares some of his personal experiences with his favorite kind of draft horses. In this first chapter, veteran Canadian Belgian breeder Eddie Freitag from Saskatchewan tells the Belgian story.*

59 "The Clydesdale Is Our Favorite" . . .

When your family has been raising Clydesdales for more than 150 years, you obviously like them a lot. An interesting look at these draft horses—known as the "gentle giants"—is presented through the eyes of Michigan breeder Don Castagnasso and his son, Tony.

66 "It's the Percheron for Us" . . .

The blacks and the greys are definitely the favorite breed for the Art Bast family. But it's interesting to see how this Wisconsin family got started with Percherons back in 1934. They have been going strong with this favorite breed of theirs ever since.

75 "The Shire Is a Beautiful Horse" . . .

This breed lost much of its earlier-day popularity, but the Shire is now making a comeback in the United States. This horse—often mistaken for the Clydesdale because of its similarities—is described through the eyes of three veteran breeders: Idaho's Sue Wilson, and Maryland breeders Norbert Behrendt and Howard Streaker, Jr.

80 "We Like the Westphalian" . . .

For all practical purposes, this German breed was practically extinct here in the states. At the low point there were only seven Westphalian horses in this country, but two Indiana draft horse breeders—Leonard and Larry Fox—are trying to get this rapidly-vanishing breed back on its feet.

85 "The Suffolk Is My Choice" . . .

A trip to England a few years back sold Bill Hardt on the merits of the Suffolk breed of draft horses. Soon after this Illinois farmer returned home, he made arrangements to import three of these horses, bred strictly for farm work, from England.

88 The Breed That Didn't Make It . . .

Ever hear of the "American Cream" draft horse? It was a breeder's dream, the only draft horse breed ever developed in the United States. But the American Cream breed is now practically extinct—today there are only a few of these horses still around.

90 Lead a Horse to Water . . .

After you've finished the close look at the various breeds of draft

© Copyright 1977 by Reiman Publications, Inc. Library of Congress Card Number 76-45044. All rights reserved. No part of this publication may be reproduced in any form or by any means without the written permission of the publisher. For further information, write: Publisher, Reiman Publications, Inc., 733 N. Van Buren, Milwaukee WI 53202. Printed in the United States of America.

Front cover photo by James Fain. Back cover photo by Frank Lessiter. Inside front and back cover photos by J. C. Allen and Son. Photos on table of contents pages by J. C. Allen and Son, Buck Miller, Dale Stierman, Jos. Schlitz Brewing Co. and Frank Lessiter.

Breeders share their ideas on seven draft horse breeds . . . . . See page 54.

Draft horse interest once again is jumping skyward. . . . . . See page 94.

When horses and teamsters together did the farm work . . . . See page 102.

When horses were used to turn the nation's furrows......See page 119.

America's sport of horse pulling is making a comeback...See page 133.

Taking big horses into the show ring was like a "disease"...See page 168.

horses in the previous six chapters, we offer you "a breather". This chapter offers you a refreshing pictorial look at horses appreciating a nice cold drink after a day's hard work.

94 The Return to Real Horsepower...

A common love for breeding and showing big horses by hobbyists and part-time farmers has sparked a comeback for draft horses in recent years. See how a school superintendent, an electrical contractor and a government worker raise three breeds of draft horses in their spare time.

102 When Horses Tilled the Land...

Early days on the farm meant plenty of hard work for both teamsters and horses. Here's a look back into the history of agriculture, showing the many early uses of horses on the farm.

119 Plowing with Horses for the Fun of It...

Competing in plowing contests has once again become a modern-day sport for teamsters. They like to show how they can still handle a good team and plow a straight furrow. It also gives them a chance to show present-day folks how horses and teamsters used to work the land.

126 Plows Have Long History...

More than 375 patents had been assigned to plows and plowing improvements by 1855. The interesting history of wooden plows, cast-iron plows and steel plows is traced through the years.

132 Get a Horse...

Some folks wrote the horse off too early when automobiles and tractors came on the scene. But when those mechanical vehicles got stuck in the mud, it was horses that often came to the rescue.

133 The Unmatched Excitement of Horse Pulling Contests...

Competitors from the word "go," a team of pulling horses love the thrill of matching efforts against a tough-to-haul load. One of Americas's great rural sports, horse pulls continue to grow.

148 A Champ Shares His Pulling Secrets...

One of the country's top "pullers," Wisconsin's Marshall Grass shares his secrets on selecting a team, getting them in shape and driving them to top honors in horse pulling contests.

154 Horse laughs...

Well, even horses sometimes need a few moments to just relax and pass the time by cracking a few jokes. For photo evidence, turn to this chapter for a little horsey humor.

156 Meet a Blue Ribbon Horseman...

Indiana's Harold Clark has probably won more blue ribbons than any horseman in the world. Handling horses in the show ring for 54 years, he's earned the reputation as one of the greatest.

168 Show Ring Fever Gets in Your Blood...

Once you start showing draft horses, you usually can't quit. Some folks refer to it as a "disease" that no medicine will cure. Justin McCarthy has spent 37 years showing horses and recalls a tough class in Chicago that included 65 two-year-old stallions.

178 Winning Combo: Draft Horses and Draft Beer

Here's the whole story of the famous Budweiser eight-horse Clydesdale hitch... from the day it was formed right up until today. Thanks to television commericals and appearances from coast to coast, these are the most famous horses in the world.

186 43 Years of Selling Draft Horses...

Arnold Hexom has sold more draft horses than any other auctioneer in the world. He once spent 25 straight hours on the auction block without ever stepping down. Yet his greatest thrill came with a six-horse hitch of matched grey Percherons.

192 When Horses Hauled the Freight...

America was built on the legs of the draft horse. Big horses did it all in the old days, keeping towns and cities on the move.

200 The Woodsman, a Saw and His Horse...

While the horse has been replaced by mechanical monsters in most areas, horses still "rein supreme" in some timber-cutting operations. Their ability to maneuver around trees and stumps keeps both horses and lumbermen in business deep in the woods.

205 Where the Horse Is Still King...

Draft horses or bicycles—that's the only transportation choice you have on Michigan's Mackinac Island. Trucks and cars have been banned from the island since 1928. So it's up to 450 draft horses to supply the island's summer "horsepower" needs.

215 Drays, Horses and Freight...

On Mackinac Island, only horses move the freight. Watching these teams at work gives observers an opportunity to see how America's freight was moved years ago.

222 When Farm Kids Horsed-Around...

It's difficult to separate a farm boy from his dog today. In the old days, it was often a boy and his horse that were friends.

230 Medicine, Harness and Horseshoes...

While teamsters had plenty of skill when it came to caring for horses, there were times when a veterinarian, harness-maker or farrier had to be called in. With the regained popularity of horses, these same specialists are again being kept busy.

238 His Goal: The World's Heaviest Horse...

Darwin Starks' ambition is to get his name in the Guinness Book of World Records... along with a draft horse, of course. He nearly did it a few years back when he tried to feed a horse to the heaviest weight in the world. Then tragedy struck...

242 Eastward Ho the Wagons: American History in Reverse...

For one group of draft horse owners, the nation's Bicentennial celebration in 1976 was more than just a chance to remember how their ancestors moved west in covered wagons. These modern-day pioneers made that same trip—only west to east.

The excitement of mixing farm kids with draft horses .....See page 222.

The farrier kept busy pounding out a living with horses.....See page 230.

Modern-day pioneers celebrated our own Bicentennial .....See page 242.

"Giddap!" (Foreward to Our Book)

UNMATCHED MEMORIES. The days are gone when draft horses were used as shown here—to haul corn and to rake hay. But the memories will never die. A team was more than just a pair of animals—there was a "kinship" between a farmer and his team. They could count on each other to farm the land. Ask any oldtimer —he'll likely remember the names of his horses better than those of his neighbors.

I GUESS I've always wanted to write this book. It's just that I never got around to it till now.

You see, since I never was able to fulfill my childhood dream of owning and driving a team of big draft horses, it seems only logical that I should use the field I chose instead—agricultural journalism— to write a book about these big "gentle giants."

My love for these handsome beasts with the flowing manes, clopping feet and rippling muscles traces back to the days when I was a grade schooler in rural Michigan. Every time one of these great teams came in sight I could feel my pulse quicken—it was my life's ambition to drive one of those big six-horse hitches when I grew up.

That looked like "the life." I guess it all started when my

Grant Heilman

J. C. Allen and Son

Grant Heilman

dad and I used to watch those big hitches at the Michigan State Fair. The drivers sitting up on those wagon seats high above our heads, in confident command of those six big horses—gosh, it made my chest ache with envy.

Unfortunately, none of those drivers ever invited me aboard. How I would have liked to climb up on that wagon seat and feel the "power" of those six lines running through my fingers. To a young farm boy like me, that would have been the greatest thrill in the world.

When You Say Budweiser...

The "king of all hitches" in those days was the six-horse hitch of Clydesdales the Budweiser folks used to bring to the Detroit-based Michigan State Fair every few years. Those big, handsome horses prancing down the street pulling the big red beer wagon was a beautiful sight for my young eyes.

I Missed a Great Deal

They always traveled in two big vans—I sometimes suspected they kept all the horses in one van and case-after-case of Budweiser in the second van.

Well, driving a six-horse hitch for a living never came to pass for me. But with my Walter Mitty imagination, I'd still like to...

Although tractors were coming in fast during the late 1940s and early 1950s when I was growing up, my father still used a team of Belgians on our Michigan farm.

Sadly, I now look back and recall my delight with the tractor...and my scorn for those draft horses. I'd drive the tractor every chance I got, and turn up my nose at taking the lines of the team. Horses just weren't fast enough for me—besides, they didn't come with a comfortable seat, hydraulic hitches and power-take-off.

Looking back, I know now that I missed a great deal by not

AN EIGHT HORSE HITCH? This may look like an eight horse hitch, but it isn't. The two horses on the far left are pulling a two-row corn planter. The other six horses are hitched to a harrow or disc that's working the soil ahead of the corn planter. The horses were probably grateful for the photographer's signal to "hold it" since it gave them a few minutes of rest on this long day during the busy spring planting season.

During a recent Michigan Draft Horse Show, I talked with Justin McCarthy of Parnell, Michigan. As he was telling me about how he had shown draft horses at the nearby Western Michigan Fair in Grand Rapids for many years, I volunteered that my grandfather and great-grandfather used to show Shorthorn beef cattle at that same show.

To my surprise, Justin remembered my grandfather and great-grandfather very well... and proved it by telling me a few things about my ancestors that were definitely fact. That's some memory, because my ancestors quit showing Shorthorns at that fair in the early *1920s*.

"Horse-Lovers" Are Nice People

Before I "get down to business" in this book, I would like to thank all of the draft horse people, farmers and many others who so willingly helped make this book possible, through their time and patience in vividly recalling memories and stories of early draft horse days, and their aid in helping me track down other individuals and photos that would help tell this important story. You've all been highly cooperative, and it's obvious horse-lovers have one thing in common—they're warm, friendly "people-lovers" as well.

There are many moments and friendships that went into this book that I will cherish for many years to come.

Frank Lessiter

being more interested in draft horses. And I'll bet there are many other folks who grew up on farms during that period who now feel the same way.

While today's farmers use big horsepower tractors and super-powerful machinery, many still fondly remember when the draft horse was *king* on their farms. Those were the days, my friend, when you talked of handling farm work with one, two, three or four horsepower—not 100, 200, 300 or 400 horsepower.

Bringing those early days of *real horsepower* into proper focus is one of the main reasons I decided to produce this book. I felt it was time someone produced a book that combines the nostalgia of the "hay days of real horsepower" with the present-day accomplishments of the great draft horses. Hopefully, a thorough book of these two phases of "horsepower" will provide a good picture of the important role draft horses played in the development of American agriculture.

A 60-Year Memory

I've met a lot of great people and enjoyed many interesting experiences while producing this book. One moment sticks out above the others since it had a special *family* meaning to me.

EDITOR'S NOTE: We think you'll find this the most complete book ever published on draft horses, at least from the <u>owner's</u> point of view. Our intent isn't to advise you on how to buy, breed, drive or manage draft horses. Rather, our aim is to bring you some of the history, feelings, experiences and the outright "love affair" draft horse people have had through owning and working with these magnificent steeds.

While the rest of this book is put together in a logical sequence—tracing the history of draft horses from the days they were <u>needed</u> to the present when owning them is mostly a nostalgic hobby—we've pulled out one chapter from that sequence and plopped it right up here in the front of the book. To us, that seems appropriate, since the ultimate excitement of driving a big hitch can be grasped in no better way than to share the experiences of Dick Sparrow, the man who drives "the 40".

Jos. Schlitz Brewing Co.

Jos. Schlitz Brewing Co.

He Drives the Most Famous Hitch of All—"The 40"

WHILE DICK SPARROW'S training as a "40" driver started when he was seven years old, he didn't know for 33 more years that he was going to actually drive 40 horses at one time. Sparrow started driving multiple-hitch teams on a two-row cultivator at the age of seven. Then he advanced to three horses on a ground tiller ahead of a corn planter and drove teams on a one-bottom sulky plow. By age 15, he was driving five and six horse hitches on a harrow.

WHAT'S IT LIKE sitting high above the crowd, holding two big fists of lines and looking over the backs of 40 big Belgian horses in your sole command?

If you're a Walter Mitty fan,

Jos. Schlitz Brewing Co.

11

ICK SPARROW has his hands full when he drives the 40-horse
tch. With 40 big Belgians out front, the hitch is a re-creation of a
rn-of-the-century circus spectacle. The Schlitz "40" was the first
ch big hitch to pound the pavement since 1904.

A PRE-HITCH MEETING let all crew members get their specific responsibilities straightened out. It took a big crew to harness, hitch and handle the 40 Belgians during the early on-farm training days.

Harold Cline

os. Schlitz Brewing Co.

it's something you dream of, but probably would never do yourself.

But it's no dream for "40" driver Dick Sparrow of Zearing, Iowa.

He got his chance to put together and drive a "40" in 1972, the first such big hitch since 1904. And it's no easy job to control these 40 light chestnut-colored sorrel Belgian mares and geldings that average a ton apiece, and stand 16-½ to 18 hands at the shoulder.

Sparrow handled the 40-horse hitch under the sponsorship of the Joseph Schlitz Brewing Company from 1972 to 1975. In later years, Sparrow contracted directly with parades, fairs and shows with sponsorship help from the O's Gold Seed Company.

Sparrow compares driving a 40-horse hitch to holding back a locomotive under a full head of steam. There's a subtle comparison there... for Sparrow, a rugged farmer who stretches 241 pounds of muscle over a six-foot tall frame, is built much like a locomotive. No Hollywood producer could have cast anyone better for the role of a big, burly

"40" driver than Sparrow.

In later years, Sparrow became the first driver to ever handle 48 horses in a hitch, doing it at the Iowa State Fair and in the Cotton Bowl Parade at Dallas, Texas.

More Than a Family Affair

But it takes much more than just Sparrow to make the "40" go. A crew of 27 men and women, along with his entire family, have played an important part in the success of the "40."

Sparrow's wife, Joy, supervises much of the behind the scenes work. Son Paul serves as his dad's backup driver. Another son,

WAS IT POSSIBLE FOR 40 HORSES to be hitched together and actually driven by one man sitting atop a big circus wagon? Hundreds of visitors sought the answer as they turned up at the Dick Sparrow farm near Zearing, Iowa, during the early-day practice sessions with the big hitch. Sparrow and his crew proved it could be done... and done much more easily than some folks had thought. It was quite a sight as the 40 horses were harnessed, hitched and driven out into a soggy field.

Harold Cline

Harold Cline

Jos. Schlitz Brewing Co.

Robert, is one of the "40's" outriders. The three Sparrow daughters, Elaine, Sue and Rose, help with the meals and other chores when not attending school. Sparrow's father, Ross, also frequently traveled with the "40."

"American kids need time to play," explains Sparrow. "But they also need a lot more time to work. That's one of my strongest beliefs." The Sparrow family gets plenty of opportunity to work, since they farm 1,100 acres and raise corn, soybeans, oats and hayland which is grazed or harvested each year.

Some 60 beef cows roam 80 acres of permanent pasture on the Sparrow farm that hasn't seen a moldboard plow in almost 50 years. Sparrow also finishes his own calves, and purchases steer calves to glean corn and alfalfa fields. It's obvious from these numbers and this workload that Sparrow is a full-time farmer in addition to his "40" activities.

In the pre "40-hitch" days, Sparrow fed 500 steers per year. But due to all of the extra work with the horses, he finally had to cut back on cattle feeding.

The Sparrows continually have 50 to 60 head of draft horses on the place to keep their "40" team fully "staffed," and these horses require the full-time efforts of four men. Two others handle the beef cattle and field work. During the peak "tour" season, when the team is in constant demand for appearances, the work force is stepped up considerably.

Horses Work, Too

But it isn't all "show" and no work for the horses. When they're home, they're called on often to help—they're used to haul manure, seed oats and for other pulling tasks. They're also used for giving weekend hayrides in summer and bobsled rides in winter.

Sparrow's training as a "40" driver started when he was seven years old, although he wouldn't know for 33 more years that he was going to actually drive 40 horses at one time.

Dick Sparrow has always had a deep love for horses—especially draft horses. He—along with his father and grandfather—was born in the farmhouse where he lives today. He learned to drive and break horses from his father and uncle.

Sparrow started driving multi-hitch teams on a two-row

HUGE HANDS enable Dick Sparrow to hold the 10 lines that control the 40 horses. Gloves protect against painful skin burns when the lines whistle through his fingers as the hitch quickly pulls out of a turn

"Whoa Now... All of Ya!"

HOLDING THE LINES

Here's how Dick Sparrow holds the 10 lines—some more than 100 feet long—in his huge hands, controlling the 40 horses in the hitch:

• Lines for the number two team go under his little finger, and between his thumb and index finger.

• Lines for the number five team go between his little and ring fingers, but over the thumb. These lie on top of the number two team lines.

• Lines for the number eight team go between his middle and ring fingers and over the thumb. They lie atop the other lines.

• Lines for the lead team go over the index finger and down the palm.

• Lines for the wheel team go between his index and middle fingers and down the palm. They lie below the lead team lines.

cultivator when he was only seven. Then he advanced to three-horse teams on a ground tiller ahead of a corn planter and on a one-bottom sulky plow.

By age 15, Sparrow was driving five-and six-horse teams on a harrow. He started driving competitively in the show ring at age 19.

Sparrow started showing draft horses in 1948, eventually walking off with grand champion ribbons at many fairs. A particularly big win was when the Sparrow farm had the champion Belgian lightweight gelding at the 1964 Royal Winter Fair in Toronto, the world's largest show of draft horses.

With his interest in multiple-horse hitches, it was a natural for Sparrow to put together a 12-horse hitch for the second Milwaukee Circus Parade sponsored by Schlitz in 1964. Sparrow continued to drive a 12-horse hitch every year in the circus parades through 1970.

Parade Needed Fresh Look

After several years of these highly successful parades, the Milwaukee Circus Committee decided something new and

---

"Driving the '40' is like trying to hold back a locomotive under a full head of steam..."

"extra" was needed to boost the already tremendous interest in the parade.

A personal passion for circus history led the late Robert A. Uihlein, Jr., the chairman and president of Schlitz and a horseman himself, to investigate the re-creation of a 40-horse hitch for the 1972 parade. Several 19th century American circuses, dating as far back as 1857, had featured 40-horse hitches in their street parades. The hitches were strictly show business in those days when the public really appreciated horsemanship.

Why had those early hitches featured 40 horses? Probably because 10 teams four abreast was a convenient arrangement.

After much discussion and planning, Dick Sparrow was selected in the winter of 1971 to organize, train and drive a 40-horse hitch for the 1972 and 1973 parades in Milwaukee.

"I'll have to admit that when I was first asked to organize and drive a 40-horse hitch I was confident I could do it," Sparrow says. "But I guess I didn't really know what I was in for."

He knew it had been done before—the last time was in 1904 by the Barnum and Bailey Circus people. Between 1857 and 1904, a few skilled teamsters single-handedly drove 40-horse hitches.

Yet these men had left no instructions for Sparrow to study. They were all drivers, not writers, and there was little or nothing in print about handling the big hitches.

One famous driver, Jake Posey, even predicted in his autobiography (in which he wrote only a few paragraphs about driving and hitching techniques) that no one would ever drive a "40" again. Posey had driven at the turn of the century for Barnum and Bailey Circus, and he predicted a "40" would never be driven again not because nobody would be skilled enough to do it, but because he felt circus parades were gone forever.

The last "40," before the Schlitz hitch appeared in the 1972 Milwaukee Circus Parade, was paraded in 1904 by Barnum and Bailey.

Perhaps Posey's flat statement spurred Sparrow on. He had two things on his side—skill acquired in years of driving six, eight, 12 and 16-horse teams, plus a few hazy photographs of old 40-horse circus hitches.

After his selection as the "40" driver, Sparrow drove a 16-horse hitch in the 1971 Milwaukee circus parade. He moved up to a 20-horse hitch for the annual Zearing Labor Day parade in 1971. In October, he tried an 18-horse hitch with horses standing two abreast.

The following winter, he spent a good deal of time studying the old "40" horse photos from the Circus World Museum at Baraboo, Wisconsin, to check out the driving techniques of earlier "40" drivers.

"I had about a dozen of those old 40-horse hitch photos," recalls Sparrow. "I used a magnifying glass to study those pictures for hours.

"I checked the way the teams were harnessed and the way the drivers held the lines. At the time, I often wondered why they

---

"To train for the '40', Sparrow called on 33 years of experience in driving horses..."

---

harnessed the horses the way they did. But after we got out and were actually driving our horses, I found there was a good reason for nearly all of the hitching ideas those early drivers had used."

Sparrow had about 20 hitch horses on the farm, so he had to buy 30 more horses. He quickly found out only about one out of every three horses would work in the big hitch. Many young horses would work well alone, but couldn't stand the regimentation of the "40."

At one early practice session at the farm, Sparrow was breaking four frisky colts into the number four team (counting forward from the wagon) on a 20-horse hitch. He fastened short lines, called "breaking straps," to the team immediately behind rather than run lines all the way back to the driver's seat. Thus the more experienced number three team controlled the number four team. This system worked fine, so he continued to use it.

Big Day Arrives

Just to discourage any "runaway" ideas the whole hitch might have, Sparrow drove his first "40" in a soggy farm field

Frank Lessiter

Jos. Schlitz Brewing Co.

MAKING A TIGHT u-shaped corner, the lead horses were often steered by outriders since they were sometimes out of sight of the men on the wagon.

CROWDS WERE always amazed at how easy a 90° turn could be made on the street. But it took plenty of teamwork to make it look smooth.

Frank Lessiter

ON THE NEXT page, seven men eased the "40" down a hill. The brakeman used a stick for leverage to hold the brake wheel and slow the eager team as Dick Sparrow pulled hard on the lines.

19

"Giddap...And Please Pass It On!"

on April 30, 1972. If the Belgians were going to try to make a run for it, he reckoned, this soft ground would tucker them out fast!

At that time, the lines to the number nine team, which stands right behind the lead team, had not been completed by the Iowa Amish harness makers. So, while Sparrow had planned to drive with 20 lines—10 in each hand, as the old photos showed turn-of-the-century drivers doing—he instead used breaking straps to fasten the number nine team to the number eight team. And then he drove with nine lines in each hand.

Driving that full hitch of 40 horses for the first time is a day Sparrow will never forget. For it was that day that Sparrow learned, with hundreds of spectators, reporters, photographers and television crews on hand, that *he could indeed drive 40 horses*. For Sparrow, it was a dream come true!

But he decided that day to do it thereafter with even fewer than 18 lines. "The weight of all those lines alone nearly pulled me off the wagon," he recalls.

Many practice sessions followed. Sparrow learned he had to restrain the number five and eight teams in turning the horses, which in terms of four abreast stretched out more than 100 feet in front of him. He also found he needed lines to the number two

22

(Continued on page 26)

THE LONGEST lines stretched more than 100 feet in front of the wagon. When the hitch was coming out of a turn, up to 20 feet of lines raced through Dick Sparrow's fingers in two seconds time.

Frank Lessiter

THE HITCH stretched out over 135 feet when the 40 horses and big circus wagon were rolling down the street.

Jos. Schlitz Brewing Co.

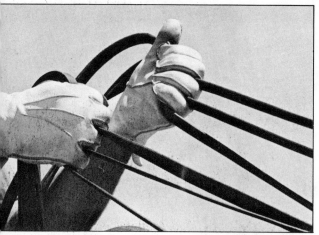

MAKING A 90° TURN with the hitch takes 20 seconds. One time the hitch turned a corner where cars on opposite sides of the street were only 16 feet apart. It was a tight fit since the hitch was 15 feet, 6 inches wide.

Jos. Schlitz Brewing Co.

Frank Lessiter

Getting the 40 Around

Since every corner of every parade route is somewhat different, it was difficult for Dick Sparrow to explain how he manages such a turn until we gave him a specific example.

Okay, Dick, let's say you're coming down a narrow 30-feet-wide street, and will turn left onto another 30-feet-wide street. This 90° corner is not absolutely square, but is rounded back about 5 feet. And just to make it interesting, let's say there is a light pole sitting right near the inside curb, so you can't run the wagon up over the curb.

Both streets are flat. (A street with a crown in the middle is a different ballgame all together, says Sparrow. But we won't get into that problem here.)

This is about as tough a corner as you want to get into. But you've made many a corner just like this one in numerous parades, Dick.

Now, tell us how you're going to make this turn with a 40-horse hitch, a big 5-ton band wagon that's 22 feet long and 12 feet high, which altogether—horses and wagon—stretch out to a distance of 135 feet? You're heading into this turn, Dick, how will you handle it?

Time to Pull Back

"When my lead team is 50 feet from the intersection and they still have 80 feet to go before they turn, I start holding back on their 100 foot lines," answers Sparrow. "I pull 'em back about 3 feet.

"Then I jerk the lines on the number eight team back about 3 feet. Now we'll move to the number five team and jerk their lines back 2 feet.

"As I start taking up the slack in the lines, the brakeman loosens up on the brake.

"We'll next go back to the lead team and pull their lines back another 3 feet. Then we'll pull back another 2 feet on the number eight team. Then pull back another 2 feet on the number five lines.

"Now, we go back to the lead team and pull another 3 feet on their lines. Then we pull another 3 feet back on the number eight team. At this point we probably need to pull the lines on the number five team back 1 foot. Then we need to pull back 6 inches on the lines of the number two team while giving the wheel team about 6 inches of slack to keep the wagon rolling.

"Every team in the hitch is now slack except for the wheelers that are right next to the wagon. Right now, we've got the lead team pulled back 9 feet and we'll pull them back another 2 feet.

"By this time, the lead team is right up to the corner. We'll probably have to take another 6 inches of slack out of the right-hand lines since the leaders are wanting to turn. We almost

24

Frank Lessiter

a Tight Corner

have to hold them back.

"Next, we have to pull the left lead line back 6 inches. Then we pull the right lead line back 6 inches. Then real quick we pull the left lead line back 6 inches which will start the leaders turning. Then you have to reach over and play out about 1 foot of slack on your right-hand line.

"By this time, you've got approximately 9 feet of slack out of the number eight team and 10 feet of slack out of the lead team. On many corners, the lead team will have disappeared from our sight by now. That's a little scary the first few times you do it.

Lines Are Crossed

"As we come around the corner, you reach over and grab about 1 foot of that right hand number five line because the lines are crossed in your hand.

"Now in the same way, we've got to get another 6 to 8 inches of slack out of the line for the right-hand number two team. And the inside left-hand wheel horse needs another 1 foot of slack.

"By this time, the lead, nine and eight teams are around the corner. But the lead team needs 5 or 6 feet of slack let out *slowly*. If you let the slack out too fast, the horses will get going too fast. And then by the time the wheel team tries to make the corner, they would be running at full speed which is dangerous.

"But if you don't let out some slack to the number eight team, they will pull you off the wagon as they pick up speed.

"About this time, the wheelers are up at the corner and starting to turn. The leaders and the number eight team are getting their heels walked on, so you just let most of the lines whistle out through your hands—not too fast or the lines will burn your hands. But you pull back on the lines of the wheelers at the same time. This lets the strain off your arms.

Lines Start to "Whistle"

"In doing this I would guess 20 feet of lines will race through my fingers in a second or two. They have to be kept straight so a twist does not knock the lines out of my hands or pull me off the wagon.

"A man sitting behind me on the wagon feeds the long English leather lines into and out of a trapdoor in the wagon top or onto the floor under the seats. He is responsible for keeping the lines straight, and that's a mighty important job.

"As the wagon starts coming out of the corner, the brakeman will tighten the brake down to slow the horses. And that's all there is to making a 90° turn with the '40'."

Time to do all this, start to finish: *20 seconds.*

team, at the end of the wagon tongue, since those were the horses that stopped the other teams. And he became keenly aware that the most important lines went to the lead team out front and the wheel team next to the wagon.

After that, they practiced with the "40" every Sunday during the months of May and June. And soon, after they had driven the "40" only a dozen times, a Universal Pictures film crew arrived to shoot a Schlitz-sponsored movie.

The entire Sparrow family remembers the 13 days the film crew was at the farm as being a time of mighty hard work. There were many mouths to feed—with all the crewmen around it was like feeding a threshing crew every day for Joy Sparrow. And Dick and his own "crew" hitched the "40" on 11 out of those 13 days. At one point, they hitched the "40" nine straight days.

And on one of those days, six members of the family hitched and unhitched the entire "40" alone because that's all the help that was on hand that day. If you've thrown a heavy harness up on a single draft horse, you have some appreciation for how much work it is to do that *40 times*, then hitch and handle all of them. And then do it all

Des Moines Register and Tribune

"No Hollywood producer could have cast anyone better for the role of a big, burly '40' driver than Dick Sparrow..."

---

over in reverse order at the end of the long day.

Sparrow eventually drove the "40" with five lines to teams one, two, five, eight and ten in each hand. And he has used that system ever since the "40" made its first major public appearance in Milwaukee on July 4, 1972.

"It's probably not as impressive with five lines, but it's a darned sight safer and easier," says Sparrow. "With the hitching method we use, we should never have any trouble. As long as I can hold the number one, number two and lead teams, I know I can control the entire '40'."

Constantly on the Road

From 1972 until 1975, the "40" made around 50 appearances in states from coast to coast. During several of those years, they were on the road better than four months.

When the "40" was at its popularity peak, Sparrow would spend six months driving 15,000 miles on the road a year. Besides buying horses, this included appearances with the hitch and looking over areas in advance of a visit by the "40."

Hitching the "40" horses usually took an hour on a hot day with a good-sized crew. Yet, once "when the chips were down," it was done in just 20 minutes.

Besides public appearances, it took the horse crew several hours

---

PARADE PROBLEMS

One of the problems encountered during the Milwaukee Circus Parade was driving the hitch over bridges with steel grating, where the horses could look down and see the river below.

Sparrow and parade officials solved this by covering the steel grating with rubber mats for the parade.

What was the longest parade route ever driven? Seven and one half miles in the Cotton Bowl Parade at Dallas.

What were the most miles ever driven by the "40" at one time? 12 miles from Sparrow's farm to Zearing, Iowa, and back again for a Labor Day Parade.

What parade lasted longest? A Rose Bowl parade. They started hitching at 4:30 a.m. and finished unhitching at 5 p.m.!

Dick Sparrow laughs when he tells the next one: "One time we forgot to hitch up four horses," he recalls. "We got almost done with the hitching before we realized we had skipped four horses. It was a lot of extra work moving several teams ahead in the hitch.

"It would have been kind of embarrassing to have gone into that parade with the 40-horse hitch and have the folks along the streets only count 36 horses."

a day to feed, water, clean and keep the horses in good shape. The horses were usually housed in their own circus tent.

Many of the crew members tired of the pace—some in a month, some in a year—so there was some turnover. "Not many people like to work all day and drive all night," explains Sparrow.

Some of the horses had to be re-shod every two weeks when on the road. Most were shod once a month.

"Yet a number of the horses weren't even shod," says Paul Sparrow. "It's a lot of hard work to keep the 52 horses we took with us shod. And their footing is often better on blacktop or concrete without shoes.

"It's easy for a horse with metal shoes to slip and fall when he hits a steel sewer cover on a street—especially on a wet, damp day. But if we felt a horse needed a shoe, we made sure he got it."

Moving the "40" was always a

THE FIRST MAN to ever drive 48 horses was an honor earned by Dick Sparrow during the 1976 Iowa State Fair. Later, Sparrow drove a 48-horse hitch in the Cotton Bowl Parade.

major undertaking. Some 52 horses traveled in four 44-foot semi-trailers. A fifth 40-foot long semi-trailer carried the five-ton, 22-foot bandwagon, the harness made by Iowa Amish craftsmen, a kitchen and other gear.

It generally took 3½ hours to

Frank Lessiter

break camp the morning of a move. But everything usually ran smoothly, since most of the crew had worked together for quite awhile.

When they were travelling, $125,000 worth of insurance was carried on the horses alone.

Can't Do It All

While Sparrow is a rugged strongman, he does not drive an entire long parade route. He alternates with his backup driver, son Paul.

"I'm simply not physically capable of driving two or three miles," explains Sparrow. "Draft horses today are bigger than they were 70 years ago, plus they're not as well broken, and they aren't as tired from other work.

"In the old days, circus horses were worked hard every day in six- and eight-horse hitches unloading the circus train. So when they paraded, they were pretty docile.

"Even if I could drive the entire parade route, I wouldn't. I think it's important that there be other people capable of driving in case I can't. If for some reason I personally couldn't make it to a parade—because of illness, a traffic jam or whatever—I wouldn't want all the people along the

AN EMBARRASSING MOMENT happened one time when the "40" almost went out with only 36 horses. Four horses were left in their stalls.

"Until I can do that—and do it repeatedly without problems—I'm never going to feel sure up there on the seat.

"It's kind of like the circus trapeze artist who is never afraid of missing the hands of the catch man...until they do the act without any kind of safety net."

His Helping Hands

In a tight situation, such as turning a narrow corner with crowds of people on both sides, two men give Sparrow a hand. Paul takes the lead team lines from his father. A man sitting directly behind Sparrow takes the wheel team lines.

On especially tight turns, 20 to 40 feet of leather lines race through Sparrow's fingers when the team straightens out again. The lines must be kept straight...a twisted line could knock all of the lines from his big hands...or pull him right off the wagon seat.

One key to Sparrow's success has been his crew of skilled outriders. "Their job is different now than it was when we started," Sparrow said. "Now I want them to do as little as possible, but to always remain alert."

Mostly, the nine outriders on saddle horses keep crowds back. But when a scary incident happened during a parade through downtown Chicago, a quick-thinking outrider saved the day.

One of the horses fell on the slick street and became entangled in its harness. It might have resulted in serious injuries, since the rest of the team could have panicked.

The team was immediately stopped. Noting how the horse was entangled in its harness, an outrider quickly sat on the horse's head. Without being able to move its head, the horse couldn't get up, which might have resulted in injury. Or the horse may have panicked in its entangled condition.

Once the harness and horse were separated, the outrider let the horse up. And the parade continued without further incident.

"That was just another example of why I've surrounded myself with people who know horses and react well," concludes Sparrow. "Driving a '40' isn't easy, but it can be done, and can be done safely."

Touch and Go

Sparrow once made a 90° corner in Mitchell, South Dakota, with only 16 feet of clearance between parked cars...and the whippletrees (the swinging bars behind each team to which the harness traces are fastened) mea-

parade route to be disappointed."

Sparrow's son, Paul, has driven a number of times without Sparrow on the seat beside him...and through some pretty tight situations at that.

"But until the day comes where I have to make a U-turn on an 85-foot wide racetrack with the 135-foot long hitch, I won't know," explains Paul.

> **ONE BIG HAPPY FAMILY**
>
> When the "40" goes on the road, it's just like one big circus family.
>
> The crew and the horses travel in style. The five semi-trailers, one straight truck, one 12-passenger personnel carrier and occasionally a car or two transport the people, horses, tent, wagon, harness, feed and gear.
>
> A sewing machine goes along to keep the crew's work and parade wardrobes in top-notch shape.
>
> A complete first-aid kit is carried to care for the crew's health. (In the first four seasons on the road with the "40", there were only six incidents which required emergency treatment at a hospital or a dentist's office.)
>
> A first-aid trunk is also carried for keeping the horses healthy.
>
> Several sets of hair clippers also are tucked in with the other gear. Both the crew and the horses get regular haircuts when they are out on the road.

29

TRAVELLING IN STYLE, the "40" included 52 horses hauled in four 44-foot long semi-trailers. A fifth semi-trailer carried the five-ton bandwagon, harness for 40 horses, kitchen and other gear. Meals were served to crew and friends at tables set up under a lean-to tent pitched alongside the kitchen trailer.

VISITORS OFTEN SAW 52 horses lined up in a ro[w] quietly eating grain. This included the "varsity 40," [a] few spares and saddle horses for the outriders. In fou[r] years, the hitch made 50 appearances. Horses we[re] housed along the way in a long ten[t.]

Frank Lessiter

Ed Schnecklo[th]

sured 15 feet, 6 inches wide.

The outriders had their horses behind the wagon and were working with the hitch horses on foot at that corner. "We would let the lead team go ahead a few feet and then wait for the rest of the horses to move ahead," recalls Sparrow.

"We just worked our way around the corner. The wheel neck tongue yoke touched a parked truck as we started to turn. Finally, the outriders had to slide a car sideways a few inches because the rear wheel of the wagon was going to touch it. The street just wasn't big enough . . . and those parked cars didn't help."

Despite unbelievable harrassment, only once have the horses ever been scared and acted like they wanted to run. (People have tossed firecrackers under the horses, shot them with straight pins, waved flags in their faces, broke balloons, cheered by the thousands and let elephants trumpet loudly in their horses' ears without any problems.)

Yet a little dog scared the horses once after the Dallas Cotton Bowl Parade by letting his paws "screech" down a screen door. "The horses tried to run, but stopped after about six feet," says Sparrow.

"Then the dog did it again and they probably ran 30 feet before we got them stopped. I wasn't holding the lines and driving at the time. But I had hold of those lines pretty fast!"

As a safety precaution, Spar-

row keeps a distance of about a half city block behind any parade unit. He feels there is no danger of the "40" running away if you use all of the safeguards that are built into the hitch and if the driver doesn't get cocky.

In fact, Sparrow believes the "40" is the safest hitch he has ever driven. He figures it is much safer than a six-horse hitch or a single horse behind a cart. Again, the old principle of "safety in numbers" applies—two horses can't run away when they're part of a 40-horse hitch.

The work of training new horses for the "40" has gone on

Frank Lessiter

THE TWO MEN who had the most to do with putting the Schiltz 40-horse hitch together were Dick Sparrow and the late Robert A. Uihlein, Jr., of the Jos. Schiltz Brewing Co. At right, a crew member grabs a fast shave before meal time rolls around.

Jos. Schlitz Brewing Co.

WHILE THE "40" TRAVELLED, farm work went on. There were as many as 500 calves to feed in some years... and horses hauled the hay. Horses handled much of the field work on the 1,100 acres of corn, beans, oats and hay the Sparrow family worked. Eight men handled the horses, cattle and crops.

WHEN THE "VARSITY 40" is home, they have to work just like everyone else. The Belgians haul hay and manure, seed oats and handle other pulling tasks. As a special treat, a few horses are used for weekend hayrides in the summer and bobsled rides during the winter.

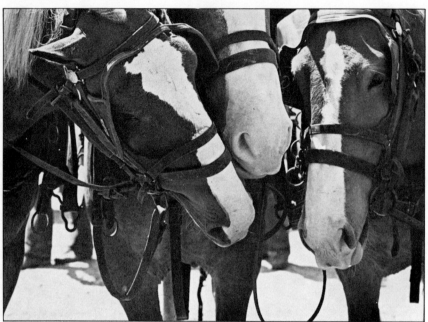

Frank Lessiter

constantly. Out of the original "varsity 40" that went to Milwaukee, only ten were still in the hitch four years later.

In training young horses for the hitch, Sparrow usually starts them in a 16- or 20-horse hitch. He will use an experienced lead and wheel team, stringing eight or ten young horses through the other teams.

"You can place about any horse in the '40' if you put him in the right spot," says Sparrow. "We've actually broken horses right in the hitch. It's a perfect place to work a young horse, but they don't learn much.

"They are like sheep. They just follow the other horses. Young horses are placed along the outside up in the center of the hitch. We don't put them in the middle of a four horse team, because our outriders couldn't get at them to calm them down if needed. And along the outside, the crowd doesn't bother them much—they feel safe in this big herd of horses.

"But you've *got* to have experienced horses as the lead and wheel teams and down the center of the hitch."

Two Favorite Horses

Sparrow named just two horses when we asked him to tell us about his favorite Belgian horses over the years.

"The most sentimental horse I ever owned was an old mare named Topsy," says Sparrow. "She knew more than I do about hitches.

"We drove her in a hitch for 21 straight years. She made a great lead horse and was in the first 12-horse hitch we ever took to the Milwaukee Circus Parade."

Okay, Dick, what was the *best* quality horse you ever had?

"I don't really know," he explains. "Maybe it is a mare called 'Sparrow's Nan.' Her hocks were off a little, so we never showed her in any of the halter classes at the fairs. But she was one heck of a horse, and still is awful close to being one of my favorites."

While she was never shown herself, three of her offspring walked away with a number of championship trophies during their show ring career.

"Sparrow's Nan" started her hitch career as a two-year-old

"When a horse fell in a parade, a quick-thinking outrider quickly sat on its head . . ."

---

lead horse on a 12-horse hitch. The year the "40" was formed, she had a young colt and was not used in the big hitch. But when Sparrow needed a spare horse to

Frank Lessiter

take to the Cotton Bowl Parade in Dallas later that year, he took her along.

"We had two lead horses get sick the night before the parade," he recalls. "So we worked this old mare as a lead horse in the parade, and even though she hadn't even had a harness on in more than a year, she did an excellent job."

Maybe Again Some Day

A couple of years later, in 1975, she started out as a wheel team horse on the "40." And later that season, she was promoted all the way from the wheel team to the lead team to take the place of a problem horse out front.

"She was one of the best and most versatile horses I ever owned," says Sparrow fondly. "She always pulled a good load, but she never wanted to pull a really big load."

Unlike Jake Posey, Sparrow doesn't say he will be the last of the 40-horse hitch drivers. "There may be another 40-horse

===
"If I hadn't had the advice of old-timers, I would have had an awful time putting this
hitch together..."
===

hitch 50 years from now," explains Sparrow. "Who knows? But even though I'm leaving more behind in photos and writing for him to learn from, that driver will likely have a hard time doing it.

"My driving of the '40' is based on a lot of personal experience and the experiences of many other men. I learned a great deal from my father and uncle who drove multiple hitches in the field and on road graders over the years. And I got help from a lot of other men who grew up with horses.

"If I would have eliminated the help and advice of men over 50 years of age who helped me, I would have had an awful hard time putting this hitch together. And that kind of advice likely won't be available to a '40' driver years from now."

Sparrow was 40 years old when he first put the big hitch together. Yet he had worked horses in the field since the age of seven. That growing-up-with-draft horses experience is something most later-day drivers won't have either.

After returning home from the 1972 Milwaukee Circus Parade in which the "40" appeared for the first time, Sparrow wondered how he had ever gotten through the parade with the big hitch. "I'm a lot smarter now about

35

driving the '40' than I was then," he explains. "And the smarter I get, the more convinced I am that we are just tapping what could be learned about driving a big hitch."

A One Act Circus

Since the "40" was put together in 1972 with a special circus flavor for the Milwaukee Circus Parade, it's appropriate to describe the Sparrow operation as a small traveling circus—with just one act. But still in the words of bygone circus days, the "40" is "the most exciting street spectacle of all time, an authentic re-creation of an American marching triumph, only many times magnified."

You don't have to talk with Sparrow very long before you realize the satisfaction he gets from the "40." "If I didn't love it, I wouldn't do it," he says. "Well, you don't like every bit of it. But you couldn't last a summer just for the money.

"There's pride in achieving something not many people could achieve. There's a great deal of satisfaction in doing any job well. Driving a 40-horse hitch stretching out 100 feet in front of you is a real challenge, just the mechanism of the thing."

While the "40" has been a lot of responsibility and problems for Dick Sparrow, the glint in his eye when he's talking about it reveals his real feelings. No doubt about it—recreating the "40" has been highly satisfying to a man who really loves draft horses.

THERE'S FENCE to be mended each time Dick Sparrow and the big hitch return home. With 60 or more Belgians constantly pushing out the barnyard fence, there's always rebuilding work to be done. Puffing on his familiar cigar, Sparrow proudly says the "40" has been a great and fascinating experience.

Frank Lessiter

$800 of Groceries Each Week

YOU GOTTA "think big" when you are handling the 40-horse hitch.

And it's true whether you are feeding the horses or the 27 men and women who travel with the group. For, wherever the 40-horse hitch goes, there are three meals a day to prepare and serve to this big, hungry crew.

Cooking for the crew—plus friends who often drop in for a meal along the way—fell on the shoulders of Sandy McBride, Zearing, Iowa, and Evelyn Lippold of Avoca, Iowa. They may feed up to 50 folks at any one meal.

Both their husbands are also members of the crew. "I got tired of staying home while my husband was on tour, so I quit my job and joined the cooking

JOY SPARROW supervises the kitchen crew that included two full-time cooks. Buying $12,800 worth of groceries during 16 weeks on the road, the cooks often served 50 crew members and friends at a single meal.

Frank Lessiter

Frank Lessiter

A COMPACT KITCHEN was built in the front of the semi-trailer that held the big wagon and harness. There is an electric stove, refrigerator, sink, plenty of cabinets, lots of overhead space and a pay telephone. Meals are served outside at tables spread out under a lean-to tent in front of the kitchen. A typical dinner for the crew would include 18 pounds of steak, 15 pounds of potatoes, 1 gallon of gravy, 10 loaves of bread, fresh tomatoes, cole slaw, milk, pop and coffee.

Frank Lessiter

crew," laughs Sandy.

Evelyn adds: "My husband had been breaking horses for years. Then he started helping Dick with the 40-horse hitch. So I decided to come along. I had cooked all my life in restaurants anyway, so this wasn't hard to adjust to."

Feeding the crew means a trip to the grocery store every day. They usually come away with more than $100 of groceries a day . . . plus they carry many non-perishable items with them in the semi-trailer that serves as their kitchen.

When the 40-horse hitch was on the road 16 weeks a year, these two gals would buy over $12,800 worth of groceries in strange, away-from-home supermarkets!

The gals' away-from-home kitchen is in the front of one of the five big, white semi-trailers used to move the hitch.

Four of the trailers haul 52 horses. The fifth trailer is used to store the five-ton circus wagon, all the harness and contains the kitchen up front. When they arrive at the parade site, the wagon and harness are unloaded, while the cooks start preparing the first meal.

The kitchen includes an electric stove, refrigerator, sink, plenty of cabinets, a pay telephone and lots of overhead storage space.

Open-Air Eating

Meals are served outside at tables spread out under a lean-to tent in front of the kitchen.

On one hot August day this writer was invited to have dinner with the crew. On this occasion, at one of the many state fairs they have visited, the two cooks and Joy Sparrow served:

• 18 pounds of round steak, cut up into individual portions and served as minute steak.
• 15 pounds of mashed potatoes.
• 1 gallon of homemade gravy.
• 10 loaves of fresh homemade bread, baked early that morning.
• Fresh tomatoes
• Cole slaw.

The menu and the scene could have been taken right out of one of the books on cooking for old-time threshing crews.

The two cooks order 10 gallons of milk and four cases of soft drinks each day. They also keep two large percolators full of coffee. Two five-gallon thermos containers are kept filled with iced tea and lemonade.

Breakfast usually requires at least six dozen eggs, toast, cereal and *plenty of pancakes*.

To save time, the crew eats from plastic plates and plastic cups. But that still leaves plenty of silverware and pans to wash after each meal. It usually takes the three women an hour to clean up.

Besides keeping the crew well-fed, the two ladies also keep them in clean clothes. Clothes are sent out to a laundry in each city.

"We once estimated that if we took the clothes to a laundromat, we would have 35 loads to do," says Joy Sparrow. "That is just too much work for us with everything else we have to get done. It would take us two or three hours a day just to do the laundry."

So Joy, Sandy and Evelyn concentrate on the cookin' . . . and do it well!

DRAFT HORSES were an essential part of life in our early-day small towns. Besides helping till the farmland, horses hauled farm familes to church and moved the freight that kept the thousands of little towns of rural America humming. Nearly every town had its own blacksmith and harness shops...businesses that have long since been replaced by automobile, truck and tractor dealerships. But in the days when the draft horse was "king," hay was the gasoline of the times.

ANIMALS FOR ALL SEASONS, draft horses were used for plowing and planting in the spring, cultivating in the summer and hauling in the crops in the fall. And on many a freezing winter day, there were always tons of manure to haul to the field. Regardless of the month of the year, draft horses never suffered from a lack of exercise. There was always farm work to be done.

H. Armstrong Roberts

Stewart Doty

When Draft Horses Ruled the Land

FOR BETTER THAN 300 years, draft horses were the main means of transporting men and goods in the new world.

- They provided power for farmers.
- They helped build railroads.
- They helped build telegraph lines.
- They helped win wars.
- They pulled stagecoaches.
- They towed canal boats.
- They transported the nation's frieght.
- They hauled the pioneers west.
- They helped build the roads that would lead to their eventual replacement—by cars, trucks and tractors.

And they did much more.

Yet few people really recognize the important part horses played in the settling and development of America.

First came the saddle horse. Without him, all travel would have had to be entirely on foot.

Next came the general purpose horse. Not only could he carry riders on his back, but he was large enough and strong enough to pull wagons, plows and other farm implements.

Then came the importation of the specialized draft horse in the mid 1850s. Our country had progressed to the point where we needed horses with the size, endurance and temperament for pulling heavier loads and bigger farm implements.

Besides working on farms, these big horses also pulled the heavy wagons that moved the

---

"Though the automobile has replaced the horse, it's a good thing for the driver to stay on the wagon..."

---

freight on the streets of our cities and towns.

So, the horse was a very essential piece of property in early America. Practically every family—whether it lived in town or in the country—owned and used one or more horses.

Hay Was "Gasoline"

Much of America was actually built around the horse—serving its needs provided business for many people. Every town had its own blacksmith shop and its own harness shop. Numerous companies were organized to manufacture horse shoes, harness, wagons, buggies, saddles and the farm implements pulled by horses.

Heavy horses were in great demand during the days when America was rapidly expanding. And hay was the gasoline of these times.

Several new developments in the 1800s brought the draft horse into its own as a means of mechanizing farm work. All hay was still scythed by hand... until the horse-drawn mowing machine was introduced in 1822. A lot of farmers wiped sweaty brows and marveled appreciatively at this fast mowing contraption.

Then Cyrus McCormick demonstrated his new horse-drawn reaper in 1831. This grain harvester—which removed the limitation on how much work a farmer could do in a day—heralded a new mechanical revolution in the fields of America.

Another significant development was introduction of John Deere's steel moldboard plow in 1837. Pulled by husky teams, it scoured the heavy Midwestern soils, and a few decades later proved ideal for breaking the matted, crusted, sunbaked soils

41

BACK AND FORTH, back and forth across the field the two horses pulled the spring tooth harrow. There was little time to rest as the fields had to be prepared as fast as possible for corn planting. It wasn't easier on the teamster either as he walked just as far in a 12-hour day's time as the horses. In fact, his feet probably hurt more at the end of the long day.

Grant Heilman

found in the Great Plains.

These developments enabled the draft horse to step to the forefront during America's first agricultural revolution. And, as men left the farm for the battlefields during the Civil War, the draft horse stepped in and filled much of the farm labor gap.

Horses Ate the Acres

This was the start of the early mechanization of American agriculture. By 1918, *26.7 million* draft horses were being used to pull the growing line of machinery that worked America's farms.

Today, draft horses have been replaced by big "acre-eating" tractors and implements that busily march across our nation's fields, doing the farmers' work faster and more efficiently. And

ANYBODY WANT TO GUESS the weight of this young horse? Weight guessing has always been a friendly rural sport—well almost always friendly. Any guessing and arguing about this animal's weight would be over just as soon as this farm woman balanced the beam on her young Clydesdale colt.

J.C. Allen and Son

SHUCKS, SOME FOLKS probably figure a rotary hoe is a fairly new piece of farm machinery. But these four horses were pulling this rotary hoe through a Purdue University soybean field many years ago.

J.C. Allen and Son

we now have less than 3 million draft horses and mules.

While most farmers in the old days plowed with just one two-horse team, there was often a neighbor who hitched several teams on one of the big gang plows. And many folks—some of

---

"No horse gets anywhere until he is harnessed . . ."
—Harry Emerson Fosdick

---

them were sons of those farmers using multi-team hitches—can still remember the tiring hours spent harnessing and unharnessing those big tandem horse hitches.

"Those hitches could be compared to a football team," recalls Rex Whitmore of Burlington, Wisconsin, who spent many an hour driving a gang plow hitch for his father. "Not all horses would work together on such a hitch. So, we had to do a lot of experimenting each spring to get six horses that would work together as a unit.

"The lead team was a long way from the plow seat. And if they weren't willing to pull hard, there wasn't much you could do about it But by hitching them differently, we could usually get a smooth working combination."

Whitmore says he was always amazed how these horses could somehow sense when the plow was about to hit a rock. "They

43

would immediately stop. And you had better believe them, too—especially if you were using a sulky plow. Nine out of ten times, they would be only inches

---

"A horse, a horse, my kingdom for a horse . . ."
—William Shakespeare

---

from a huge boulder. I still don't know to this day how they did it," he adds.

While horses may have made things much easier on the arms of the farmers who previously had to do much more hand labor, it didn't make things easier on all parts of his anatomy.

Remember those cast-iron seats on the plow or other implements? After a full week of riding one of those seats in the spring, many a young man figured he had "J. I. Case" or some other trademark branded on his backside for life!

Draft horses were the mainstay for farm power in those early days . . . until World War II came along. The farm labor

State Historical Society of Wisconsin

C. P. Fox

A QUICK ROLL on the ground is all that's needed to quickly cure almost any itchy back. Many big horses liked to scratch their back this way at the end of a hard day's work just as soon as the harness was taken off and they were led through the gate to the evening exercise lot.

MOST FOLKS would guess this is a young Shetland pony and a big Percheron horse. But some experts in the draft horse field would argue that the big horse is not a Percheron in this photo made before the start of World War I at Iowa State University. They recall that Iowa animal scientists in those days were working with grey Shire and grey Clydesdale horses. This horse is one or the other of these two breeds and not a grey Percheron, maintain some old-timers in the horse field.

STAYING CLOSE TO MOM, this young colt is about to take a "big" step into some deeper water in this northern Wisconsin river. But mom doesn't seem the least bit concerned as the two of them head out for some early morning grazing.

DINNER TIME for these young foals on the M. C. Hodgson farm at Ottawa, Illinois, usually meant a "full-feed" of ground oats and corn.

J.C. Allen and Son

United States Department of Agriculture

United States Department of Agriculture

J.C. Allen and Son

shortage during that war did as much as anything to bring the demise of the draft horse, since it forced a new spurt of mechanization. And farmers sold draft horses in record numbers as the

> "Your foot will never feel better as long as a horse is standing on it . . ."

post-war mechanization movement set in.

As tractors quickly took over the work of agriculture, many folks figured the draft horse would soon be gone forever.

Hundreds of draft horse breeders quit the business, assuming that horse prices would never again be high enough to

A FEW LOUD WARNINGS to stay away or just a friendly nip on the shoulder? Nobody knows for sure. But horses could be just as playful in the barnyard as a couple of kids out for recess at school.

A POWERFUL TEAM, these Percherons were being harnessed by Elmer Taft for a day of spring field work at Lynnwood Farm near Carmel, Indiana.

J.C. Allen and Son

NOON TIME meant dinner for more than just the farm crew. About 11:30, many a young colt would listen to its empty stomach churn with hunger as it waited for the team and men. While the farm crew headed for the big farmhouse, a colt would get his dinner, too. A colt could guzzle a fair amount of milk in a half hour—enough to keep him happy to supper time.

J.C. Allen and Son

ANY FRISKY STALLION would always run and run when turned out each morning on the exerciser cable. Running back and forth, the stallions at Iowa State University were no exception as they beat a permanent runway with their heavy hoofs under the exercise cable in this barnyard.

J.C. Allen and Son

"A GOOD BROOD MARE in harness" was the simple caption line written on the back of this old-time print. And who could argue... since she looks like the kind of working mare any farmer would have been proud to find a stall for in his barn.

48

United States Department of Agriculture

BASKING IN WARM SUNLIGHT on an early spring morning seems like the thing to do for these two young colts. After a rather chilly night, the warm rays of sunlight feel good as the new day starts.

show a profit. Many county, state and national fairs soon dropped draft horse classes or cut back on the amount of premium money being paid.

Tractors Were Menaces

Those were definitely some hard years for farmers and breeders who stayed with the draft horse instead of the tractor.

Many old-time horsemen actually resented the mechanization of farming. "The tractor was not a bad machine in its place," says Arnold Hilde of Stanwood, Washington. "But when horse machinery was taken off the market as a way to replace horses with tractors, it became a menace."

As his horse machinery wore out, Hilde was unable to buy either new horse machinery or spare parts for the horse machinery he already owned. Rather than sell his draft horses and convert to mechanized agri-

---

"Nature gives everybody five senses: touch, taste, sight, smell and hearing. But the other two—horse sense and common sense—have to be acquired..."

---

culture, Hilde switched his diversified farming operation to a livestock grazing program. (Even today he uses Percheron horses to handle two hay mowers, a side-delivery rake, a horse-drawn hay baler, a wagon and a manure spreader.)

Things Started Looking Up

The pessimism of the 1940s and 1950s gave way to optimism and renewed interest in draft horses during the 1960s and 1970s. Horse prices began to inch upward. Bigger crowds started turning out for draft horse shows. And people with a renewed interest in "real horsepower" started buying draft horses—both as a hobby and as a business.

The result is that draft horse prices have increased sharply in recent years. Part of the reason is

49

SHE'S BIGGER than the split rail fence, yet the gentleness of this Percheron mare is probably all that's needed to keep her in the pasture. But there's no doubt—she could crash through this fence on a second's notice with her tremendous power. And if she wanted to, she could probably just as easily jump right over this kind of fence that Abraham Lincoln made famous so many years ago.

that sentiment over-rules "being practical" when the huge animals are bought and sold today.

"When you worked horses in the fields in the old days, you could only afford to pay so much for a good horse," explains Bob Robinson, a Percheron breeder from Richland, Michigan. "But these days when you are raising draft horses, really love them and don't have to make a living with them, I guess

---

"In the choice of a horse and a wife, a man must please himself—ignoring the opinion and advice of friends..."
—George John Whyte-Melville

---

there's no limit to what a person will pay for a good horse."

Some folks figured the big draft horse would be extinct by now. But due to increased popularity during recent years, it's not even an "endangered species" these days. Fact is, the draft horse is making an amazing comeback.

Many of today's farmers can proudly recall how they lived during the transition from real horsepower to gasoline horsepower. They've been privileged to enjoy the best of both worlds ...and have had many memorable moments as a result.

J.C. Allen and Son

Memories of Real Horsepower

I'M OLD ENOUGH to still remember back when we farmed part of the home farm in Michigan with horses.

While we were pretty well mechanized with tractors by the time I came along, we still used one team to do some farm work.

The most horses that were ever used on our family farming operation, which goes back to 1853, were three teams of horses and a driving horse. There was probably a good reason—we only had seven horse stalls in our big dairy barn.

When I was a kid, we had a team of big Belgians named Silver and Scout. I guess it was okay for us to borrow the names of the horses that the Lone Ranger and Tonto rode to radio and television fame, since the originator of the Lone Ranger series, Brace Beamer, lived only three miles from our farm. And that's the truth.

Known for Slow Pace

Our team of Belgians was the domain of our hired man at the time, Harry Robertson. I'd swear those two horses took on the personality and movements of old Harry. They moved the same way he did—slow but steady.

Frank Lessiter

The team had an ornery streak in them, too—they were constantly opening the barnyard gate at the worst possible times.

How well I remember how my dad hated to hear the clop-clop of their hoofs out on the blacktop road in the middle of the night. Especially on cold winter nights. He used all kinds of colorful language while chasing them back to the barn and closing the gate.

TAKING A SAW to the wooden tongues, John Lessiter (the author's father) quickly switched several implements on his Michigan farm from horsepower to tractor power.

But I'll give Silver and Scout credit for one thing. They were smart. On the few occasions when my dad picked up the lines, they immediately perked up their ears and were ready to move.

But Dad was normally willing to let old Harry plod along at a slow pace with the horses pulling the hay wagon back to the field after we unloaded on a hot summer day—I'm sure it was because it gave us all a chance to rest for a few minutes.

That is, he was willing to grant that lull in haying if the field was fairly close to the barn. *And* if it didn't look like rain.

But when we were working in one of the back fields or if it looked like the hay would soon be wet, Dad would grab the lines and Silver and Scout would quickly be off at a fast trot.

We used horses in my early

SILVER AND SCOUT always knew when my dad rather than the hired man had the lines. When the "boss" was hollering "giddap," these big Belgians were smart enough to move out in a hurry. If they didn't, they knew the lines would soon be slapped across their rumps.

BY THE YEAR 1890, all farm machinery that would require horses was already invented. Many a piece of horse-drawn machinery was "put out to pasture" as tractors came in with a rush.

Frank Lessiter

P. Fox

days for mowing, raking and hauling hay, hauling manure, pulling the grain binder, cultivating . . . and pulling the tractors and our truck out of the mud.

Time Came to Change

For years, we mowed and raked all our hay with horses. But there came a time when Dad finally realized this was slowing down the haying season too much.

I still remember the day when

"Imagine that, a horse being allergic to hay!"

---

he took a saw to the tongue of the hay mower and cut it in half. After that, we mowed with the tractor. But, as with a lot of the other early "horse machinery", it wasn't easily converted to tractor use—I spent many summer days riding the mower seat to raise the sickle bar at the end of the hay field or when it plugged up with hay.

Silver and Scout were then used mainly to rake and haul hay. That was a nice slow job suited to both the horses, since they were slowing a bit, too.

Horse Allergic to Hay

But then one day Scout started coughing and wheezing while raking hay. Raking wasn't such a tedious job . . . and should have been a task any horse could handle.

After a close inspection, our veterinarian said old Scout was, of all things, allergic to hay. Now that's really something—an animal who eats and works mainly with hay being allergic to it. (Maybe old Scout was smarter than we thought!)

Since this happened during the peak of one summer's haying season, the easy solution was to get out the saw again and shorten the wooden tongue on the

---

"When my dad picked up the lines, the team always knew it and perked up their ears . . ."

---

rake. After all, just one horse on the rake wasn't very efficient.

And that's just what we did. So hay raking quickly became a tractor power job on our place. It wasn't long after that we said goodbye to our team of Belgians, as Dad sold them—with some reluctance, I'm sure—to another farmer.

Rather than being allergic to hay, I still believe old Scout was mainly allergic to work.

COMING IN SIX COLORS, most Belgians are bay, chestnut or roan colored. But you can sometimes find a few Belgian horses that are brown, grey or black.

EVEN TODAY, the Belgian is the only breed of draft horse bred to any extent in Belgium. While a few Belgian horses were imported into the United States in the last half of the 19th century, it was not until the early 1900s that large numbers of Belgian horses found their way by ship to our shores.

Frank Lessiter

Edward Druck

"Belgian Is My Breed"

WHILE THE BELGIAN is without doubt his preference, Eddie Freitag is quick to tell you why he loves these big, beautiful bay, chestnut and roan horses.

"I'd say 90% of my preference for the Belgians is simply because I was born and raised with them," says the Alameda, Saskatchewan rancher. "You just naturally prefer what you grow up with."

While the Belgian is the "Number One" breed for this western Canada cattleman and his family, Freitag is quick to recognize there's no one favorite breed for all draft horsemen. "If you are going to be a good horse breeder or livestock man, you have to appreciate all of the breeds," he adds, sounding like a true diplomat.

Always Been Belgians

Freitag has worked with Belgian horses all of his life. He started as a youngster working with Belgian horses on his father's western Canada ranch. Then he spent 12 years with Belgians at the world-famous Meadow Brook Farms in Michigan.

Next came work with Belgians at J. M. McKeehan Farms in Indiana. And finally, a few years ago, Freitag made the complete circle by moving back to the Belgian breeding he started out with as a youngster on his family's farming operation.

The home ranch, where Freitag grew up and which he now manages, includes 800 acres of prairie land. Besides raising wheat and Belgians, the ranch runs 40 Hereford brood cows.

While it's hard to pin Freitag

EDDIE FREITAG has spent a lifetime successfully showing Belgian horses. Draft horse shows and Belgian horses are a way of life for this Canadian.

Frank Lessiter

ON SHOW DAY, Freitag rolls out of bed even earlier than usual. You'll find him putting the finishing touches to hoofs while other horsemen are still curled up in sleeping bags in the barn.

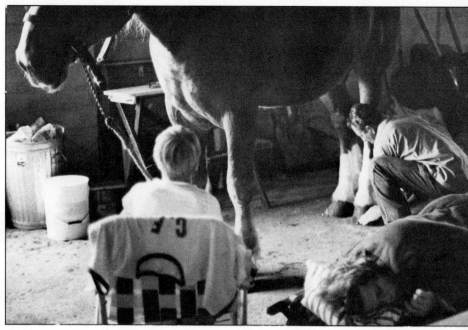
Frank Lessiter

down on just why he prefers the Belgian breed, there's no doubt it's his favorite. "I guess it's because the Belgian is an attractive horse—I like its color and form, and I like working with them," he says. "And it's still a challenge to get six horses in a hitch that look alike, have the same basic conformation and the good disposition to work together successfully."

The European Flavor

As you might have guessed, the Belgian breed originated in Belgium. It is the only breed of draft horses bred to any extent in Belgium.

As far as is known, no oriental horse breeding was ever mixed

---

"One year I traveled by car, airplane, bus and train to get to the first fair..."

---

with the native Belgian stock. Even today, Belgian horses more nearly resemble the Flemish horse than does any other draft horse breed.

In 1886, the Belgian Draft Horse Society was organized in Belgium to encourage the breeding of native draft horses. An annual show held each June since that time has attracted as many as 1,000 Belgian horse entries during some years. In addition, the Belgium government helped promote breeding of these draft horses over the years.

A few Belgian horses were imported into the United States in the last half of the 1800s. But it was not until the beginning of the twentieth century that large numbers of Belgians were brought over.

Most Popular Draft Horse

While the introduction of the Belgian breed occurred more recently than that of the Percheron, Clydesdale or Shire, the Belgian is now the most popular draft horse in America.

The Belgian shares honors with the Shire as being the heaviest of all draft horse breeds. Mature stallions weighing over a ton are common. A mature stallion will stand over 16¼ hands tall when measured at the shoulder (16¼ hands would equal 5 feet, 5 inches).

While most Belgian horses are bay, chestnut or roan in color, you may also find a few that are brown, gray or black.

Early Show Ring Start

Since his family showed Belgian horses at a number of Canadian fairs each year, Freitag got an early start in the horse showing business. His father showed Belgians for *27 consecutive years* at the Toronto Royal Winter Fair—a *3,400 mile* round-trip journey!

"We would be gone a whole month from the time we left home until we got back," recalls Freitag. "We'd always go on a special livestock train that was put together for western Canada ranchers heading for the Toronto show with their various kinds of livestock.

"I remember going for the first time to Toronto in 1950... and that turned out to be a record trip. We made the 1,700 mile trip in 69 hours. Of course, that record has been broken since that time. But it was really something to set a record on the first trip to Toronto I ever made."

Freitag says his dad always liked to take a full show string of Belgians to the fairs... especially when they were shipping by rail. "I don't know why, but we always seemed to travel with 13 horses," says Freitag. "We would use a 12-stall palace rail-

---

"I guess I just like the Belgian's color and form..."

---

road car and always put one young colt in the alley."

Toronto Show Led to Job

Freitag's first Toronto Show back in 1950 was quite an experience for a 14-year-old boy ...and it eventually led to a

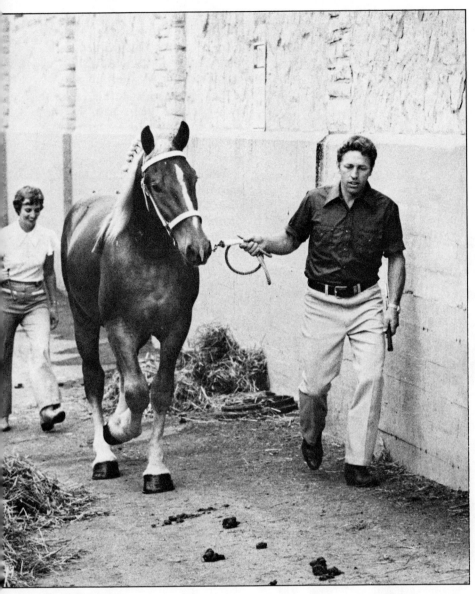

Frank Lessiter

BACK IN 1950, Freitag's first Toronto Royal Winter Fair led to a super learning experience that lasted a dozen years with the world-famous Belgians at Meadow Brook Farms. Now Freitag, his wife and their children have made the trip back to the Saskatchewan ranch where Eddie started out with Belgians when he was a youngster.

you could learn a great deal from Harold," says Freitag. "Harold was really dedicated to the draft horse business. He never stopped thinking about how to best prepare a horse for a show."

When Freitag first went to

"In 1886, the Belgian Draft Horse Society was organized to encourage breeding the native draft horses..."

Meadow Brook as a teen-ager, he lived with the Clarks. Some years he would work in Michigan during the winter. Other years, he went back to the family ranch in western Canada for the winter.

During those years when he want back to Canada, Freitag helped his dad show Belgians at a few early summer fairs. Then he would journey to Meadow Brook in time for most of the show season that included four state fairs, the National Belgium Show, the big Saginaw County Fair in Michigan, the Toronto Royal Winter Fair and the International Livestock Exposition in Chicago.

Playing Catch-Up

It wasn't always easy to catch up with the Meadow Brook show string if they had already left home before Freitag finished showing horses for his dad in western Canada. "I remember one year traveling by car, airplane, bus and train to get to Harold's first fair," laughs Freitag. "That was some experience ...and a lengthy trip, too."

After Meadow Brook, Freitag spent three years operating the

special job for him. While at that first show, he earned extra spending money by helping Harold Clark exercise the famous Meadow Brook Farms Belgian horses. This was a part-time job that was to pay handsome dividends in the years to come.

"In the winter of 1951, Harold Clark called my dad looking for a man to help with the horses in Michigan," says Freitag. "After some long arguments, my dad and mom decided I could go back to Rochester, Michigan, to work for Harold.

"I was only 15 years old at the time. And I only weighed 98 pounds.

"I think Harold had forgotten how small I really was. Because he seemed really surprised at my pint-size when I showed up. We later laughed many a time about that."

Freitag spent 12 years helping

"My father showed Belgians for 27 straight years at the Toronto Royal Winter Fair..."

Harold Clark with the Meadow Brook Belgians...and learning a great deal about the draft horse business. (This book includes a chapter on Harold Clark's 54 years of showing draft horses.)

"If you were a good listener,

57

Frank Lessiter

Belgian horse division of McKeehan Farms at Greencastle, Indiana. He continued to show at many of the big Belgian shows.

But after his father passed away several years ago, Freitag chose to return to western Canada to run the family farm operation.

While managing the family's western Canada farm operation means Freitag has to concentrate on more than just Belgian horses, the big horses still remain dear to his heart. And you can bet Canadian Belgian competitors will be seeing a lot of him in the show ring each year.

"Yes, we'll continue to show our Belgians," he concludes. "After all, the Belgians are an important part of our life."

HITCHED OR UNHITCHED, Belgian horses are a favorite of Eddie Freitag. Here he enters the show ring with a unicorn hitch.

"The Clydesdale Is Our Favorite"

WHEN YOUR FAMILY has owned and bred Clydesdale draft horses for over 150 years, it's easy to see why this "high stepping draft horse" is likely to be your favorite breed.

That's the feeling of Tony Castagnasso, who is part of the Don Castagnasso & Sons Clydesdale breeding operation at Charlotte, Michigan. "Once you and your ancestors have concentrated on one breed for a century and a half, Clydesdales just sort of get in your blood," says Tony.

Family Had Three Breeds

Tony is the fourth generation of his family to breed Clydesdale draft horses. Over the years, the family has raised Clydesdales in California, Utah, Ontario and Michigan.

Tony's great-grandfather used Belgian, Percheron, Clydesdale and several other draft horse

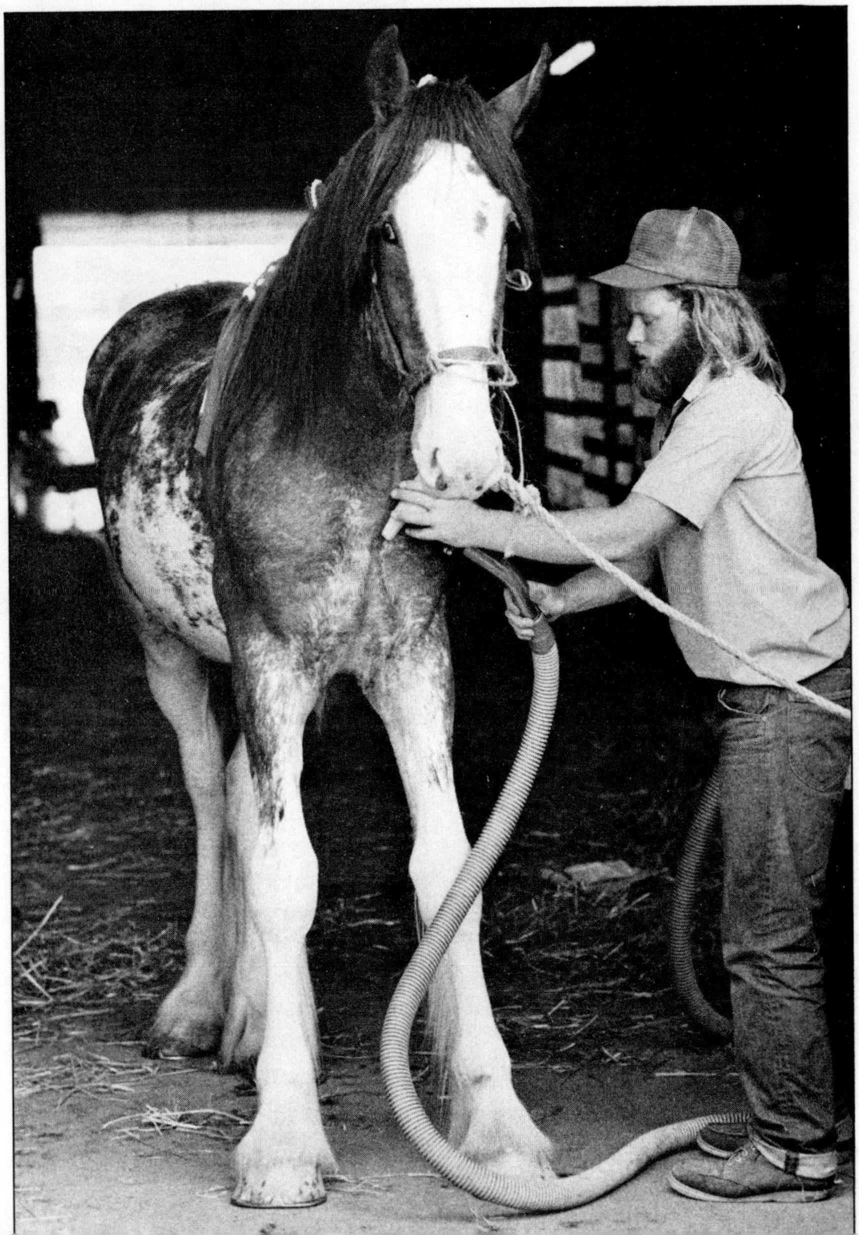

FOR 150 YEARS, the Don Castagnasso family has been raising Clydesdale horses. Unlike most owners, the horses don't do any field work on their Michigan farm.

Frank Lessiter

breeds in his California gravel hauling business.

"The more loads of gravel he hauled in a day, the more money he made," explains Tony. "Since the Clydesdale horse was a little longer-legged, he found it could take a longer stride and pull the wagons faster than the other breeds.

"The Belgians and Percherons—which had more muscle—could pull more weight than the Clydesdales. But over a full day's work, the Clydesdales could still haul more gravel.

"After seeing what the Clydesdale did in the gravel business, my grandfather and the rest of our family became more interested in these horses known as the 'gentle giants'."

Tony and his dad now operate a 200-acre farm that features a

"No other breed can even come close to keeping up with the Clydesdale at a brisk trot..."

---

few beef cows and calves along with 20 Clydesdale horses.

Unlike some draft horse breeders, they rely 100% on tractors to handle the farm work. "Our horses are primarily for showing, so we don't do a bit of farm work with our Clydesdales," says Tony. "If you work horses in the fields, we've learned they don't pick up their legs as high when we put them in a show hitch. Plus, their heads may droop when they work in a hitch.

"They simply aren't as lively when they go into a hitch if they have been working at home."

A Personal Preference

While Don and Tony both definitely prefer the Clydesdale draft horse, they readily admit their own personal preference simply came from growing up with these big horses. Still, like many other draft horse breeders, they appreciate any good draft horse regardless of the breed.

"Yet I've gotten along better with Clydesdale horses than I

Frank Lessiter

have with other breeds," adds Tony. "The Clydesdale seems to learn quickly, is quiet, very gentle and can be easily trained."

Don says the Clydesdale has always been popular in ranching country. "If you were plowing, you needed a stouter, stronger pulling horse—such as the Belgian or Percheron—than if you were mowing hay," he explains. "That's why the long-striding Clydesdales were more popular in cow-calf grass-feeding areas than in areas where you had to work your horses hard all of the time."

Tony adds that the Clydesdale really performs best in hauling work. "No other breed can even come close to keeping up with the Clydesdale at a brisk trot," he adds. "But a Clydesdale won't get down and pull real hard.

A COOL SUMMER SIP awaits this foal. Tony Castagnasso says the foals accompany their mothers to the fairs and shows. Traveling along to these events is a real learning experience for the young foals. It lets them get used to loading, trucking, people and the whole show ring ritual.

MOST CLYDESDALE BLOODLINES run about the equivalent of two human generations, says Don Castagnasso, shown decorating a mare's mane on show day. He tries to always keep several different bloodlines going in the herd's home-bred stallions.

Frank Lessiter

Frank Lessiter

SILKY FEATHERED FEET on the Clydesdale take plenty of work to keep clean. Many farmers in the old days disliked using the Clydesdale for field work since it was hard to keep the long white hair clean. There was no problem when Clydesdale horses were used to haul freight on hard surface roads. Clydesdale breeding today encourages even more feathered hair on the legs.

61

Maybe he's just too smart for that."

Breed Traces to 1715

The Clydesdale hails from Scotland. And it has been practically the only draft horse used in that country for better than 250 years.

While the early history of the Clydesdale breed is quite obscure, the breed definitely had a mixed origin. During the formation and early stages of development of the breed, both Flemish and English horses were extensively used. The Clydesdale breed can trace its beginnings back to around 1715 on farms located near the river Clyde.

The modern history of the breed began around 1750. Scottish farmers in the Lanarkshire area (the old name for Lanark-

---

"A Clydesdale won't get down and pull hard. He just may be too smart for that . . ."

---

shire was Clydesdale) began to use Flemish stallions on their native mares. The goal was to increase the size and improve the quality of their work horses.

Besides farm work, the Clydesdale breed was developed for work in the local coal fields and for heavy hauling on the streets of Glasgow.

By 1830, Lanarkshire farmers and other farmers in near-by areas had developed a system of district stallion hiring. Continued to this day, the system has done more than any other single thing to distribute good Clydesdale sires. This allows a number of Clydesdale mares from several farmers to be bred to one outstanding stallion.

There are two other important milestones in the development of the breed. One was the start of the Glasgow Stallion Show around 1870. This famous show is still held each year.

The second milestone was the organization of the Clydesdale Horse Society of Great Britain

Frank Lessiter

THE EYE OF THE judge carefully looking over one of the young Clydesdale horses led by Tony Castagnasso belongs to Wreford Hewson of Beeton, Ontario. Hewson, among the top Clydesdale breeders in the world, has more than 100 Clydesdale horses on his Canadian farm.

HIGH LIFTING FEET are a characteristic of the Clydesdale. Each foot at every step will be lifted high enough so you can see the inside of the horseshoe.

Frank Lessiter

Frank Lessiter

THE SNAPPY CLYDESDALE trot with hocks well flexed is a favorite of spectators everywhere. For style and action, Don and Tony Castagnasso feel it is hard to beat the Clydesdale.

and Ireland in 1877. This was followed two years later by formation of the forerunner to the present Clydesdale Breeders Association of the United States.

The first Clydesdales brought to North America were probably imported into Canada by Scottish farmers who had settled there. By the 1870s, large numbers of Clydesdales were being imported into the United States both through Canada and directly from Scotland.

Known for Style and Action

No other draft horse breed equals the Clydesdale for style and action. The prompt walk with a good, long, snappy stride—as well as a sharp trot with hocks

63

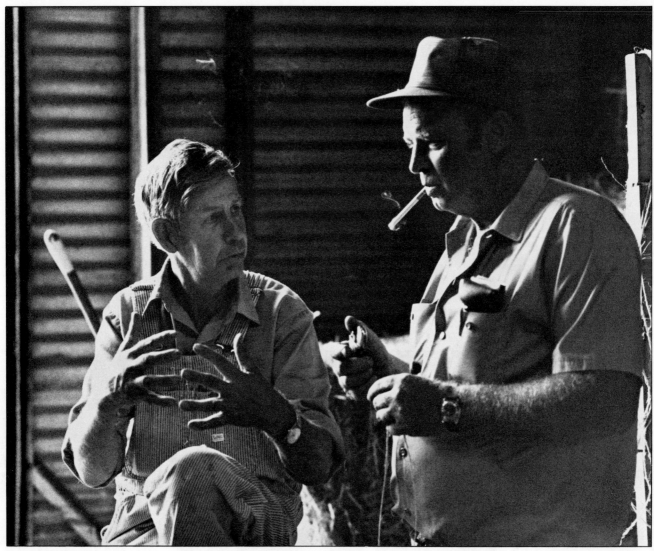

DON CASTAGNASSO lets fellow Clydesdale breeder Myron Avery, Marshall, Michigan, make a point about his stallions.

well flexed and carried close together—are noted characteristics of the breed.

Large round feet, a well-set, fairly long sloping pastern, plus a moderate amount of fine "feather" (long hair) at the rear of each leg below the knee and hock are other well-known features of these horses.

Early-day criticisms of the breed by farmers dealt with the lack of body size, the lack of width, the lack of depth, the abundance of feather, and too much white coloring with no regular distribution. Many farmers disliked the white face and white legs of the Clydesdale, since they were hard to keep clean.

The reason the long feathering white hair on the legs was not popular among farmers was that it took extra care to keep the

"My great-grandfather found he could haul more gravel in a day's time with Clydesdales..."

legs clean when the horses were worked in muddy fields or along dirt roads.

The most common Clydesdale colors are bay or brown with white markings. Yet black, gray, chestnut and roan Clydesdales are occasionally seen. But you seldom see a bay or brown Clydesdale without a white face and white feet and legs.

Several Bloodlines Going

The Castagnasso family tries to keep several different bloodlines going in the home-bred stallions they use in their herd. They try to keep 10 mares—each with a slightly different bloodline—busy raising foals each year.

Most Clydesdale bloodlines run about the equivalent of two human generations. While draft horse styles change slowly, the breed has undergone some distinct changes in recent years.

The major bloodlines currently being used in the Castagnasso Clydesdales happened pretty much by accident.

In 1962, Don went back to the source, so to speak—he made a shopping trip for Clydesdale horses to Scotland. He purchased a mare, but left her in Scotland until after she foaled. She was to be shipped later to the Castagnasso farm.

Unfortunately, the mare died during foaling. After communicating with Don, the Scottish farm-

> "Today's trend is toward more white feathered hair on the legs —especially if the horses are shown or used in parades..."

er shipped another mare and her 6-week-old foal to the Castagnasso farm.

"We raised the foal and have really liked her bloodlines," says Tony. "We have kept four of her offspring and have sold several other foals from her.

"It was just an accident that the first mare died and that we got the second mare and foal instead. While we had been working on another breeding line for five or six years, the breeding line which resulted from this accident has been one of our most successful breeding lines."

One of the major aims in breeding Clydesdale horses today is to keep plenty of hair on the legs. In the early days of the breed, this feathering was pretty scarce. Plus, the hair was purposely kept short when the horses were worked in the field.

But this leg hair is now regarded as highly attractive, and today's trend is toward more hair on the legs—especially if these horses are to be shown or used in parades.

"We oil the hair to keep it soft so it won't break off and will grow longer," says Tony. "The Clydesdale is a big horse, so most problems are in the feet and legs."

Two old Scottish sayings illustrate the value of feet and legs in development of the Clydesdale breed: "No foot, no horse" and "Tops may go, but bottoms never."

Tough, But Oh, So Gentle

Commonly known as "the gentle giants" in Scotland, the Clydesdale won this label because of its disposition, and the way it moves with the greatest of ease...yet with a powerful style. All points of the Clydesdale's feet hit the ground evenly while its long, sloping pasterns serve as shock absorbers.

The front legs and feet move out squarely from the body with a free, smooth motion and go high enough to attract the attention of all observers. The foot at every step will be lifted high enough and cocked back far enough so you can see the inside of each horseshoe.

This high stepping movement—along with the white stockings and feathered hair to match—is one of the reasons the Clydesdale has been so popular and successful in the big show hitches. During a 22-year period from 1913 to 1926, Clydesdale six-horse hitches won 21 blue ribbons at Chicago's International Livestock Exposition.

Hit Nine Shows a Year

The Castagnasso family usually shows horses at state fairs in Michigan, Wisconsin, Indiana and Ohio, takes in the big horse show at Toronto, exhibits at the annual Michigan State Draft Horse Show and goes to three Michigan county fairs in a year's time.

Besides showing in the various halter classes, they exhibit their Clydesdales in the single cart, single tandem, unicorn, two-, four- and six-horse hitch classes.

Tony says the unicorn and single tandem classes are more popular in Canada. Canadian shows also include a class for a single horse hitched to a wagon instead of a cart.

The Castagnassos took part in even more draft horse shows when they lived in Canada. "We made 35 one-day draft horse shows one year when we lived in Canada," recalls Tony. "Some other farmers made as many as *50* of these shows with their draft horses in a year's time.

"Many were Saturday shows. You could load your horses in the early morning, drive to the fair, show your horses all day and come home that same night and do chores."

Most foals go with the mares to the shows when summertime rolls around. The men work carefully with the foals from the time they are two days old.

"Going to the fairs is a valuable experience for foals," says Tony. "They learn how to load in a truck, how to get along with people and quite a bit about the show ring ritual."

Foals are harnessed for the first time at two years of age. Some geldings work their way

> "Our best bloodlines were actually the result of an untimely accident..."

into the hitch by the time they are three years old. By the time they reach four years of age, these horses are real pros in the hitch.

"After hitching a Clydesdale gelding only five or six times he will know what he is doing," adds Tony. "We haven't had much trouble with our hitch horses."

These hitch horses are driven with two bits in their mouth. One is the traditional bit that is used for steering.

The other is a "check bit" to hold the horses' heads up so they look better in a four- or six-horse hitch. Hooking on the Scotch collars, these check bits can be used to "balance out" the heads of an uneven pair of horses.

The Castagnassos have obviously learned all the "tricks" of showing Clydesdales. But then, after 150 years of owning, breeding and showing these "gentle giants", they should.

LARGE NUMBERS of these horses were imported into the United States starting in the 1870s. This was before the horses were officially recognized by France as the Percheron breed in 1883.

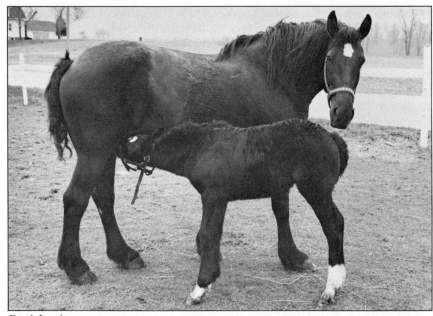

Frank Lessiter

ON THE MOVE, Jon and Art Bast roll their four horse hitch of matched black Percherons into the ring at the Wisconsin State Fair. While 90% of today's Percherons are black or grey, a few bay, brown, chestnut, or roan colored horses show up in the breed from time to time.

Frank Lessiter

"It's the Percheron for Us"

PUTTING TOGETHER a herd of quality draft horses can't be done overnight. When Art Bast and his family started raising Percheron horses in 1932, they never dreamed the breed would do so well. "But we really didn't develop top-quality Percherons until the early 1960s," recalls Bast. "Since then, we've had many enjoyable times with them."

THE BLACKS AND the grays are definitely the favorite draft horses of the Art Bast family.

It's not that they have anything against any of the other draft horse breeds. It's just that the Hartford, Wisconsin, family got started with Percheron horses, did some interesting line-breeding that turned out some outstanding horses, and has been "sold" on the black and gray horses ever since 1934.

The entire Bast family is involved in the Percheron operation. Besides Art, there are his wife, Hazel, and their three children . . . Muriel, Jeanne and Jon. And they *all* work with the horses.

"We had regular draft horses up until the early 1930's on my dad's farm, and none of them were very fancy," recalls Bast. "We did use a few Percheron horses earlier than 1934, but we'd never raised any Percheron foals."

In those days, the family farm operated under the name of Val Bast and Sons. Besides the father, Val, there were three sons in the partnership: Ray, Roland and Art, whom we interviewed for this chapter.

Interestingly enough, each of these three sons eventually went into farming on his own. And each of the three sons eventually earned considerable renown as a draft horse breeder.

Started with a Stallion

Art remembers that his father, plus he and his brother, got

Frank Lessiter

67

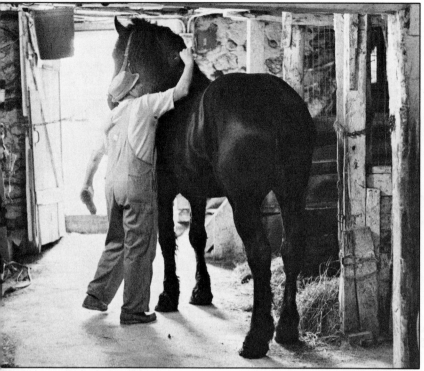

IN THE EARLY 1940s, horses were hardly worth anything. Bast remembers how breeders would ask you to take a horse "free" just so it would have a good place to live.

Frank Lessiter

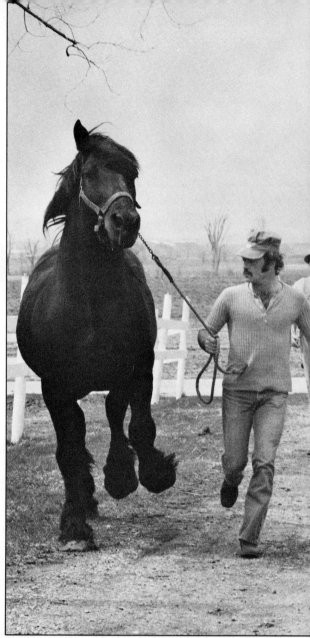

ONE OF THE BEST Percheron sires the Bast family has ever used gets a bit of exercise.

Frank Lessiter

serious about Percheron horses when they bought their first black stallion in 1932. "We wanted a good stallion to mate to our mares," says Bast. "We just happened onto a good Percheron stud that was available in the area."

Once the Bast men saw the good looking colts sired by this stallion, they bought some purebred Percheron mares. Soon after, they jointly decided to concentrate on breeding only Percheron horses.

Sitting at the kitchen table today, Bast can count off a dozen or more reasons why he prefers the Percheron over all other draft horse breeds. "But basically, once you have the bloodlines, you want to continue with them," he says. "This is true of any breeder regardless of the breed or even the species of livestock he raises.

"I've always liked the blacks and grays. So we've stayed with our favorite Percherons."

Breed Hails from France

The history of the Percheron horses stretches back 1,250 years...even before the advent of the breed. The breed is based primarily on Flemish horse

---

"I've always liked the blacks and grays. So we've always stayed with our Percherons..."

---

breeding...with several rather liberal mixes of Arab horse breeding over the years.

The predecessor of the Percheron breed can be traced back to 732 A.D., when the Moors from North Africa were defeated at Tours and Poitiers, France. The stallions that the defeated Moors rode into battle fell into the hands of French farmers and were promptly distributed to farmers throughout that area of France.

These stallions—made up of Arab, Barb and Turk breeding—provided the first mix of Oriental horse breeding into the basic Flemish-bred horse that eventually became known as the Percheron.

A second input of Oriental horse breeding occurred during the 12th, 13th and 14th cen-

Frank Lessiter

OLD MOLLY came along with the Bast family when they moved to their own farm in 1942. The old registered Percheron raised 14 colts and later became the grandmother of one of the farm's best sires.

Frank Lessiter

NEARLY EVERYONE had started replacing horses with tractors in the 1940s when farm help got scarce due to World War II.

turies. This was when the successful Crusaders brought back more Arab stallions from Palestine as "spoils of war."

French farmers continued to breed these horses with their mixture of Flemish and Oriental breeding. Efforts were soon made to develop this mixture of horse blood into a true type draft horse that was soon to become known as the Percheron.

Starting in the 1870s, large numbers of these horses were exported to the United States. This was done even before the horses were officially recognized as the Percheron breed since it was 1883 before the Percheron Society of France was formed.

So you can now see how the Percheron horse has changed greatly from its forerunner. The horse that started out as a great steed for soldiers during early-day battles has evolved over the centuries into the present-day draft animal.

While 90% of today's Percheron horses are black or gray, you occasionally find a few bay, brown, chestnut or roan colored horses in the breed.

A mature Percheron stallion stands 17 to 18 hands tall and weighs 1,900 to 2,100 pounds. While Belgian and Shire horses

"History behind the Percheron horse stretches back some 1,250 years..."

are larger, the Percheron is bigger than the Clydesdale or Suffolk.

Time to Go It Alone

About eight years after the Bast family started breeding Percherons, Art decided it was time to farm on his own. Leaving the father-son partnership, he and his wife of two years brought part of the Percheron herd to their present 83-acre farm in 1942. This farm had been owned by Art's father-in-law.

One of the horses they brought was an old registered mare named Molly. Art was unaware at the time of the strong influence she would have on the type of Percheron horses they would produce in the future.

"She raised 14 colts, had a good pedigree, enjoyed outstanding conformation and later became the grandmother of one of our best sires," recalls Bast.

Dairy-wise, the early years of farming on their own were pretty good for Art's family. But the early 1940s wasn't a good time to be breeding and selling draft horses. World War II was in full force, farm labor was scarce and farmers were starting to mechanize. Horses of all types were being sold at a record pace.

"Draft horses weren't worth much of anything in those years," says Art. "A good mare would bring only $100 to $125.

69

PLENTY OF EXERCISE from the time horses are young colts is essential to keep their legs and feet in good condition. The Bast family has the best luck by pasturing stallions the year-around.

STARTING ON A SLEIGH, horses on the Bast farm are broken to work in the winter months. Next comes the spreader in the spring. Then comes summer work on the hay mower and wagon.

Frank Lessiter

Frank Lessiter

TWISTING THE LINES around a piece of lumber will almost always keep the team from walking off.

A FEW GOOD BLOODLINES lets Bast line breed his horses. While it works, Bast recognizes you must be careful when close line-breeding.

Breeders would actually ask you to take some good horses off their hands at no cost just to give them a good home.

"Everyone was using tractors in the 1940s when farm help got scarce due to World War II. Even the farmer who could have still used a team or two decided to mechanize to keep up with his neighbors who were using tractors."

The Percheron horses on the Bast dairy farm did all of the field work in those days. Besides milking a good-sized dairy herd, the Bast family raised 125 acres of hay, small grain, corn and peas.

Calves, Cows, Colts, Mares

Bast kept milking cows until 1960. He still remembers how a

"Things got so tough one year that I drove taxi just to keep some food on the table. I'll never forget that winter . . ."

---

neighboring ex-farmer talked him into selling off his dairy herd.

"This neighbor used to milk our cows when we were showing horses at the fairs," remembers Bast. "He was always after me to come and work in the dairy bulk tank manufacturing plant in Hartford.

"When things really got tough in the dairy business, I did sell off the cows. I figured if I got rid of those cows—which were losing me bundles of money at the time—then I could concentrate on breeding Percheron horses. So we sold the cows and I went to work in the factory to support my family."

But it didn't work out. Just a year after Bast started work in the plant, he was laid off. And soon after, the firm moved the plant 100 miles west. The Bast family didn't want to move from

Frank Lessiter

Frank Lessiter

THE THRILL OF COMPETITION— and the question of whether a blue ribbon will be won or not—is at hand as Jon Bast takes the family's Percheron hitch into the ring to do "battle" with a dozen other teams.

WITHIN DAYS OF BIRTH, the foals are worked with by the Bast family. Later training is made much easier by working with foals from a very young age.

Frank Lessiter

the farm, so Art had to look for other work.

Drove Taxi, Drove Horses

"I'll never forget that winter," says Bast. "I had to drive a taxi in Hartford just to keep some food on the table."

By the next spring, he had landed a job as assistant manager of the county's Dairy Herd Improvement Association. He moved up to manager in 1962, a job he has held ever since.

Art, Hazel and son Jon supervise 200 dairy herds that test 12,000 to 13,000 cows each month. With all of the morning and evening milk testing that is done by these three, it's unusual when the family can sit down together at the kitchen table for breakfast or supper. Noon-time is about the only time you'll find them all together.

1960s Brought Better Prices

Over a cup of coffee now, Art vividly recalls those early days in the Percheron business. Depressed draft horse prices continued for nearly 20 years until the early 1960s rolled around. Despite the fact that many draft horse breeders felt the horse was doomed forever, prices started coming back to respectable levels as the 1950s came to a close.

"It was about then that people started looking for good horses

---

"During the 1940s and 1950s, a good mare would only bring $100 to $125 . . ."

---

and were willing to pay a good price for them," says Bast. "These were mostly younger guys who hadn't grown up with horses. They were the ones getting into the draft horse business in the 1960s and 1970s."

Horse prices have continued to rise in recent years. Today, it costs a "pretty penny" to buy a good stallion or mare regardless of the breed you choose.

Ribbons Everywhere

Showing their Percheron horses during the summer and fall at various fairs has been a Bast family tradition for many years. It started in 1936 with one colt, and has developed to the point where the family may take as many as 10 horses to six or eight draft horse shows in a year's time. This includes the local county fair, several state fairs, the Toronto Royal Winter Show and one or two other big shows.

Take a look around their farm home and you see a lot of evidence that the Bast horses have done well over the years at the big shows. The house is filled with ribbons and trophies that were won by their horses in individual halter and in hitch classes.

Unlike many other draft horse breeders, Bast has always preferred to "line breed" his Percheron horses. He likes to seek out a few good bloodlines, then keep them going by constantly breeding back to them. He'll tell you that some of his best colts have resulted from half-brother and half-sister matings.

THE 2:30 A.M. HORSE SALE

EARLY one morning a few years ago everyone at the Art Bast home was sound asleep when the telephone rang. Checking the time, Art found it was only 2:30 a.m.

"The man on the other end—who I didn't know—wanted to know if our sorrel stud colt was still for sale," recalls Bast. "Only half awake, I told him it was."

After haggling over the price awhile, the man agreed to buy the colt, right there on the phone. He said he would send a check and pick up the colt next week.

"So I went back to bed," says Bast. "As I'm dozing off, the telephone rings again. It's now 2:45 a.m. I'm now talking to a breeder who had looked over this same sorrel colt five months earlier.

"He wants to buy that colt. But I have to tell him I just sold the colt 15 minutes earlier.

"Well, there's all kinds of screaming and hollering on the other end of the telephone. Then this breeder tells me, 'That darned fool went and bought your colt, didn't he?'

"I didn't know what he was talking about."

Later, Bast learned both men were playing in a lengthy middle-of-the-night card game. The breeder who had been to the Bast farm to see the colt got to bragging after a few drinks about the good stud colt he was going to buy.

"So the first guy excused himself from the card game for a few minutes, called me at 2:30 a.m. and bought the colt as a practical joke," says Bast. "Then he must have said something to the other farmer back at the card table that made him wonder whether the colt had been sold in the past five months. So he called me 15 minutes later to buy the colt. But it was too late."

After all the stomping and hollering was over, there was a happy ending. The practical joker sold the colt ten days later to the breeder who really wanted it. But Art doesn't know how much the joker marked up the price . . .

"You can get poor horses out of a good sire if you aren't careful . . . and that isn't good," says Bast. "So you have to be careful which animals you use from a linebred mating. But we've found it pays not to jump around and buy a new stallion every two years.

"Breeding horses is a lot like breeding dairy cows. The farmer who breeds 36 dairy cows to 36 different bulls doesn't get anywhere. But the guy who breeds all his cows to only two or three bulls can make some real progress. And breeding to just one or two good stallions is how you make progress in the draft horse business."

Bast figures you have to be

Frank Lessiter

AT THE END of another 17-hour day, Art Bast lugs the harness down the alleyway past the stalls. Hanging the harness on a peg near the door of the barn, he'll be sure the Percherons have grain and water before heading to the house.

"fussy" about feet and legs when it comes to putting together a sound draft horse breeding program. Without sound feet and legs, a horse will never be a big success.

Most Interesting Moment

Bast says the most interesting thing that has ever happened to him in the draft horse business dealt with the sale of three mares a few years back.

Bast had sold his quota of Percheron horses for the season, and didn't plan to sell anymore when a Japanese dealer came looking for Percheron horses. So he put an exorbitant price tag on three mares, and to Art's amazement, the dealer bought all three mares.

"While that was one of our most profitable moments, it was one of the saddest days of my life when I saw those three mares being loaded on an airplane at Chicago's O'Hare airport for the trip to Japan," says Bast. "Those were three valuable mares that

---

"By selling our colts young, we can give our remaining horses plenty of individual attention..."

---

we could really have used. I was just sick to see them go. But I'm sure the three mares are making a good contribution to Percheron horse breeding in Japan."

Bast has no idea where the three mares ended up in Japan. "We sold them to a horse dealer," says Bast. "He wouldn't tell us where the mares were going since he didn't want U.S. breeders dealing directly with the Japanese farmers in the future. He intended to protect his value as a middleman in any future draft horse sales."

Lots of Fun

When they started with Percheron horses, the Bast family never dreamed they would do so well. They didn't get into the business with profits in mind; they got into it for the sheer fun of it.

"We really didn't develop good Percheron horses until the early 1960s," concludes Bast. "But since that time, we've had a great many pleasant times with our Percheron horses."

"The Shire Is a Beautiful Horse"

A RESURGENCE of interest in the Shire draft horse breed is underway. While there was plenty of early-day interest in the breed here in the states, the breed, for all practical purposes, had practically died out over the previous few decades.

But interest in the breed has been rekindled in the past few years. And, as a few horsemen took a fresh look at the breed, Shire numbers started gradually moving back up.

Sue Wilson, who now keeps 45 Shire horses at her ranch near Pingree, Idaho, originally started breeding Shires in Maine. She later moved her Shires to Idaho since it was a better area for raising heavy horses.

"I needed to find an area more conducive to breeding horses than the wet climate of Maine,"

BREEDERS SAY they can't turn out Shire foals fast enough to keep pace with the growing demand for these beautiful horses. The Shire is often mistaken for the Clydesdale.

Frank Lessiter

A FEW BREWERIES in England still use draft horses to deliver their special brew to neighborhood taverns. One south London brewery, Young's of Wandsworth, still delivers beer on short hauls cheaper with these Shire horses than by truck.

THE VERY LEGGY, tall English Shire type of horse on which breeders are trying to develop a bigger foot is part of the wide range of type found in Shire horses. There is also the native Shire that is shorter-legged, heavier-muscled and smaller-footed with less feather on the leg.

British Airways

Frank Lessiter

MORE ELEGANT than some other breeds is the way some folks describe the Shire. Some people starting out with draft horses like the Shire because it is showy and has the white feather on the leg.

she explains. "Idaho provides a better growing season. All around, it's better for the horses. The horses thrive on the dry grasses and native forage out here."

A "True Breeder" of Shires

Studying for a master's degree in biology, Miss Wilson is using her scientific background to further develop the Shire breed. A student of both the breed and

76

Frank Lessiter

Miss Wilson imported her first Shire horses from England in 1970. A major problem in importing is the many weeks the horses need to acclimate themselves to their new home.

"Importing doesn't present all that many problems as long as you know the sellers," she explains. "The biggest problem is coordinating, especially when you bring over a group of horses that came from a number of different English farms. You can't get them together in a hurry."

Miss Wilson believes the Shire horse is again coming into its own. The Shire, like all other draft horse breeds, is growing in numbers and popularity as affluent people get interested in raising big horses during their leisure time. "Still, we need some additional people interested in improving and maintaining the

"The National Brewery eight-horse Shire hitch made 274 appearances during its three-year existence . . ."

pureness and quality of the breed," concludes Miss Wilson. "And this support is coming—slowly but surely."

Over the past five years, Shire breeders have gone from registering around 10 horses a year up to about 60 animals annually.

Massive, Wide, Long Horse

While the origin of the breed is hard to trace, and what we were able to assemble is more or less speculative, it is generally agreed that the early Shire resulted from very mixed horse breeding on English farms. First developed for military purposes, the breed was soon being adopted for farm and freight hauling work.

British Shire breeders organized the English Cart Horse Society in 1878. The name was changed 6 years later to the Shire Horse Society.

The Shire is a massive horse with a wide, deep and long body.

modern breeding programs, she has a distinct dislike for "hack" breeders who are only interested in making a fast buck instead of improving Shire breeding stock.

"I really like Shires, because they are a good 'mix' of horse breeding," says Miss Wilson. "They are real showy, yet make good work horses. I think they are the best mixture of work and pleasure to be found in any of the draft horse breeds."

Breeders say Shires are very responsive, docile, determined and easily controllable. While the horses are powerful, they still manage to retain their gracefulness.

"Shires are tremendously beautiful," adds Miss Wilson. "Yet they don't need all of the care the other 'showy' breeds do."

FOR THREE YEARS, Baltimore's National Brewing Company had an eight-horse black Shire hitch. Howard Streaker of West Friendship, Maryland, operated the hitch.

Frank Lessiter

Frank Lessiter

GROWING APPEAL for the Shire has been its value as both a work horse and a show horse.

Shire stallions routinely weigh better than 2,000 pounds.

While the Shire is more rangy than the Belgian, it stands taller than any other breed. Stallions

"Shires are the best mix of work and pleasure to be found in any of the draft horse breeds . . ."

standing 17 hands or more in height are commonly found.

Heavy bone and feather are also characteristics of the Shire breed. While the more common colors are bay and brown with white markings, occasionally blacks, greys, chestnuts and roans are seen.

While there are many similarities between Shire and Clydesdale horses, the two breeds are very distinct.

The Shire is more massive and heavier-bodied than the Clydes-

Frank Lessiter

SHIRES CAN BE bay, brown, black, grey, chestnut or roan. This black colt was foaled by the old grey Shire mare you see here.

dale. And the Shire's feather (the long hair on the legs) is more abundant and coarser than the Clydesdale's.

Shires were first imported into the United States in 1853. The

"The Shire horse was originally thought of in England as a military horse..."

American Shire Horse Studbook shows a small number of stallions imported in 1880... and more than 400 imported in 1887.

Started with Black Shires

Back in Maryland, a pair of veteran horsemen are also doing all they can to develop the Shire breed. Norbert Behrendt of Highland, Maryland, and Howard Streaker, Jr., of West Friendship, Maryland, imported a Shire stallion from England in 1967.

Since that time, they have imported several more Shire horses. They are now expanding their own home-bred Shire horses based on importations of foundation breeding stock from England.

Besides breeding Shire horses, Streaker operated an eight-horse black Shire hitch for Baltimore's National Brewing Co. from 1971 to 1973. During that three-year span, the hitch appeared in 274 parades, demonstrations and exhibits.

After Streaker first got the go-ahead to organize the brewery hitch, he journeyed to England to buy Shire horses. He bought eight Shires on his first trip, then went back a few years later for four more Shires.

The hitch featured black Shire horses with a blaze face and four white legs. "At that point in time, there were probably only 600 Shire horses in all of England," recalls Streaker.

"So it was quite a job finding just what we wanted. There were lots of greys, lots of bays and lots of blacks with two or three white legs. But there were very few black Shires with four white legs to be found anywhere. We finally found them though."

Streaker says the supply of Shire horses in the United States can't keep up with the demand. People now getting into the Shire

"Most rewarding is the personal contact with these sensible, lovable giants..."

business are folks who grew up on farms and then moved away. But now that they've made some money, they are buying a few acres in the country and are again interested in heavy horses.

Easy Chair Introduction

Behrendt fondly recalls how he first became interested in Shire horses. "About a dozen years ago, I was sitting in an easy chair one night leisurely looking through a book on horses," he recalls. "I happened to see some Shire horses pictured from the state of Washington.

"After writing to the owner, I learned there were hardly any Shire horses in the United States. I soon found myself more interested in Shires. It wasn't long before I was really hooked on the breed."

But it was 1967 before Behrendt finally purchased his first Shire horse. For nine years, he and Streaker concentrated on breeding Shires of all colors. But they now emphasize bay Shires since most horses used in exhibitions and parades are bays.

"Perhaps the most rewarding aspect of owning Shires is the personal contact with these sensible, lovable giants," concludes Behrendt. "There is also a deep satisfaction gained from assisting in the preservation of a true friend of man. The Shire horse is definitely that."

WHILE WESTPHALIAN HORSES are still exhibited in their native Germany, the horses are losing out as German farmers switch to more mechanized horsepower.

Westphalian Horse Society of Germany

NEARLY 5 TONS of Westphalian horseflesh are on the move at this German draft horse show. Westphalian stallions, weighing around 2,000 pounds each, are known for their tremendous strength and agility.

Westphalian Horse Society of Germany

"We Like the Westphalian"

"LENZ! DODI! BARTEL! Peter! Hansel! and Woerf!"

Those aren't the usual names you would call out if you wanted six draft horses to come up to the barn.

Yet those were the names of the six German-born Westphalian draft horses brought to the United States in 1965. They eventually became the Meister Brau Brewery's six-horse stallion hitch.

The horses—which hailed from the Westphalian area of northeastern Germany around Munster—were originally brought to the states for the Bavarian beer garden that was part of the 1965 New York World's Fair. Meister Brau officials acquired the horses during the fair, purchasing them from the German Trade Minister and pledging that the horses would never be separated from one another.

These original Westphalian horses became the cornerstone around which the Meister Brau hitch was developed. Three more stallions and two mares were brought to the Meister Brau farm in Illinois from northern Germany in 1966.

Before these five additional horses were purchased, their ancestry was closely examined over several generations. Brewery officials wanted to be sure all

THE YOUNG SON of Leonard and Virginia Fox got interested in draft horses at an early age. The 30-pound boy and 2,000-pound Westphalian stallion keep an eye on each other.

Frank Lessiter

future foals would match the other Westphalian horses already in the states.

Finally two more stallions were brought from Germany by Meister Brau after a stopover at Expo '67 in Montreal, Canada.

Many Years of History

The matched "Westphalische Kaltblut" stallions in the hitch represent a horse breed with a long and colorful history. They are descendants of steeds that carried heavily-armored knights charging into battle during medieval times.

The massive Westphalian stallions, weighing over 2,000

> "There are two types of Westphalian horses—the cold bloods and the hot bloods..."

pounds each, were known for their tremendous strength and agility. These horses were later used to pull heavy beer wagons over cobblestone streets and along muddy roads as Germany became more progressive.

Hot Bloods, Cold Bloods

There are two types of Westphalian horses today—the "hot bloods" and the "cold bloods". The cold bloods are the draft horse version of this famous German breed. Yet they are few in number today.

The Westphalian draft horse is definitely on the decrease in Germany. With German agriculture greatly mechanized, farmers are relying less and less on real horsepower to farm their land.

The hot blood Westphalian horses are a cavalry type horse still used in the sport of jumping and for police work in Germany. The majority of the Westphalian horses in Germany today are of the hot blood variety.

The German government is doing its part to increase the numbers of this breed. For example, the German government keeps Westphalian stallions of both types in their major stallion stations. Each spring, these horses

Frank Lessiter

are placed in small stallion stations throughout the country so farmers can bring in mares for breeding. This gives German farmers the opportunity to breed mares to a top quality Westphalian stallion without having to actually own a stud.

Hitch Fell on Lean Days

Meister Brau featured the Westphalian horses in a six-horse and four-horse hitch for several years. Eventually, however, the brewery went bankrupt, and the horses, wagon, equipment and trucks found their way to the Pioneer Park at Aurora, Illinois. They remained at this park for several years... until the hitch eventually became too big for the park to financially handle.

That's when Leonard and Larry Fox stepped into the Westphalian picture. They purchased all the equipment and moved the horses to Leonard's farm at Crown Point, Indiana. By this time, the Westphalian horses were down to three mares, two geldings and two stallions.

"My husband has always raised horses," says Virginia Fox. "And his brother raises Percherons.

"The two of them figured they might be able to have some

TO SAVE THE BREED, the Fox family rescued the seven remaining Westphalian horses. Moving them to their northern Indiana farm, they hope to raise some foals to keep the breed going here in the United States.

NEVER SEPARATE THEM said the German Trade Minister when he agreed to sell the Westphalian horses to the Meister Brau Brewery in 1965. The horses have all been kept together since that time despite three different owners.

Frank Lessiter

Frank Lessiter

Frank Lessiter

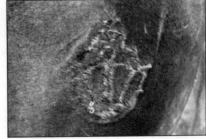

A GERMAN CROWN was branded on the left rear flank of each of the imported Westphalian horses.

OFTEN MISTAKEN for older-style Belgians, the Westphalian horses are short, chunky and very wide.

83

FANCY HARNESS for the Meister Brau hitch was just as beautiful as the horses themselves. The red and silver-trimmed harness was handmade in Germany especially for this hitch at a cost of $18,000.

Frank Lessiter

fun trying to keep the Westphalian breed going. So we bought the horses."

Looks Like Older Belgians

Mrs. Fox thinks many horsemen could easily mistake the Westphalian horses for the older style Belgian draft horse. "While the Westphalian horses are short and chunky, they are a very wide horse," explains Mrs. Fox. "The width of these Westphalian horses is really unbelievable. They also have those beautiful heads and eyes."

Was a Stallion Hitch

In the early days, the Meister Brau hitch consisted of six stallions. They later switched to a four-horse hitch and often times used geldings in the hitch.

"I saw the six-horse Meister Brau hitch once in Chicago's Amphitheater" recalls Mrs. Fox. "I noticed they had two drivers up on the wagon seat handling the lines. One handled two horses while the other driver handled the other four horses.

"One man just couldn't handle six stallions all the time. Stallions in a hitch would probably be fine most of the time—but all of a sudden they might get ornery."

The Fox family is convinced there is a place for the Westphalian draft horses in the states. And they hope to see the breed grow in the years to come.

NINE SMALL BELLS hang on the harness for each of the six horses. So you can imagine the beautiful music that came from the jingling of 54 bells as the hitch made its way down the street during a parade.

Frank Lessiter

"The Suffolk Is My Choice"

A MORE ENTHUSED horse breeder than Bill Hardt would be hard to find.

The Litchfield, Illinois, farmer is enthused about draft horses, all right... the Suffolk breed of draft horses in particular.

One of the reasons for his unequaled enthusiasm is that he is helping bring back a draft horse breed thought to be on its last legs not too many years back.

At the breed's low point, Hardt estimates there were fewer than 50 registered Suffolk horses to be found anywhere in the United States. But now that a renewed interest in the breed has taken hold, there are better than 200 of these chestnut-colored horses in the world.

The bulk of those 200 Suffolk horses are about evenly divided between the United States and England. A few are also found in South Africa and several other countries. And Hardt has personally seen nearly all of those Suffolk horses during several tours of horse breeding farms in this country and abroad.

Liked What He Saw

For example, in 1971, Hardt travelled overseas with other draft horsemen to take a closer look at the heavy horses being raised on European farms The group looked at Percherons in France, Belgians in Belgium and Clydesdales Shires and Suffolks in England.

As a direct result of that trip, Hardt imported two Suffolk mares and a Suffolk stallion later that year from England.

Soon after, he returned to England for another look at

C.L. Marley

SUFFOLK NUMBERS have increased greatly since Bill Hardt imported a pair of mares and a stallion from England.

85

C.L. Marley

THE ONLY HORSE ever bred strictly for farm work is how the Suffolk is known. But one fault is the breed's lack of size.

Suffolk horses. This time he travelled over 1,900 miles through the English country in search of the chestnut-colored horses.

The native home of the Suffolk breed is Suffolk county in eastern England. Most of the breeding of this fine draft horse in the early days was confined to this area of England.

Some breed authorities believe the Suffolk horse originated around 1700, possibly as a descendant of Normandy horse stock.

Yet the breed's foundation is

"I guess you have to say I am a draft horse nut . . ."

usually traced back to a prolific chestnut-colored stallion known as the "Crisp Horse" that was foaled in Sussex, England, in 1768. This stallion is credited with being the ancestor of all stock registered in the English and American Suffolk horse studbooks.

The first volume of the Suffolk studbook was published in 1880, even though horsemen in the East Anglia district of England had already kept private records on their horses for many years.

Suffolk horses were first imported into the United States in the early 1880s. Only small numbers have been imported at any one time since, possibly because of the Suffolk's lack of size when compared with other draft horse breeds.

Another reason that more horses were not imported was probably due to the fact that they were not bred in very large numbers in England. There was always considerable demand for these horses by English farmers.

Bred for the Furrow

Hardt and other breeders of Suffolk horses say this is the only draft horse breed that has been bred exclusively for farm work. No Suffolk horses were ever bred to handle city dray work.

For generations, the horses were bred for strength, stamina, docility, longevity and soundness—all important to the farmer who works his fields with draft horses.

Hardt says the Suffolk exhibits a ready willingness to work,

"At one time, there were fewer than 50 Suffolk horses in the states..."

---

great endurance and has the remarkable quality known as "heart".

"They are fast walkers and easy keepers, all factors of great importance to early-day farmers," adds Hardt. "These same qualities, plus gentleness and great intelligence, make Suffolks popular with teamsters and pullers."

Breed Almost Died

Like all draft horse breeds in North America, the Suffolk was hit hard by the big push to mechanize farms in the post World War II period. Actually the Suffolk breed was hit harder than most of the other draft horse breeds.

While the breed had made great strides in the late 1930s, the Suffolk breed did not have the necessary number of horses to withstand the onslaught of mechanization in the late 1940s and the very early 1950s.

For a few years in the 1950s and 1960s, the Suffolk Association even ceased to function. But then came the 1960s, and the draft horse market began to recover.

Soon after, the few remaining widely-scattered breeders who still had faith in the Suffolk heavy horse took steps to reorganize the association. With some outstanding importations of Suf-

---

"The Suffolk is the only draft horse bred strictly for the furrow..."

---

folk horses in the early 1970s, the Suffolk breed is once again on the move.

As you talk with Hardt, you quickly realize few men have such dedication for anything as this Illinois farmer has for the Suffolk horse. He readily admits, "I guess you could say I'm kind of a draft horse nut!"

You soon wonder how he developed this intense interest in draft horses...and Suffolk draft horses in particular. And when you talk with him awhile, the story behind it all slowly starts to unfold.

With a background as an Illionis dirt farmer, Hardt has owned draft horses and mules for nearly 50 years. He bought his first draft horse when he was 15 years old.

While Hardt purchased his first tractor in 1948, he still did most of the field work on his 160-acre farm with horses and mules as long as he could. Hardt still helps out on the 720-acre farm operated by his two sons, Karl and Edwin.

Nice Chestnut Color

Back when he was doing all of his farming with *real* horsepower, Hardt bred, raised and worked a great many mules. His favorite mule color was always sorrel—the result of breeding jacks to sorrel mares.

"I guess the fact that Suffolks reproduce only in this similar chestnut color was a big attraction to this breed for me," he explains. "I do know that the Suffolk is the only major draft horse that breeds completely true to color."

Chestnut Only Color

Out of better than 12,000 Suffolk matings carefully investigated over many years, not a single foal of any color but chestnut was ever discovered.

However, there are seven shades of the chestnut color. These range from a dark liver color to a light golden sorrel color. White markings occur on Suffolk horses, but they are not usually as prominent as with other draft horse breeds. Most of the white marks are confined to a star or "snip" on the head, or to white ankles or fetlocks.

"This uniformity of color makes it easy to mate a pair of Suffolk horses into a good-looking team or hitch," says Hardt. "We have found the chestnut color is very popular and readily saleable."

The Future Looks Good

Bill Hardt sees a promising future for all draft horses—and for the Suffolk heavy horse in particular. "While I don't see any great return to horse farming, draft horses will continue to serve the purposes of many people," he points out.

"No doubt the Amish farmers

"The growing popularity of the draft horse—and hopefully the chestnut-colored Suffolk—is here to stay for years to come..."

---

will continue using horses, as will some gardeners, loggers, cattlemen and others. Shows, parades and hobby interests are also on the increase."

So the day of growing popularity for draft horses—and hopefully, as far as Bill Hardt is concerned, the great Suffolk horse—appears here to stay.

—*By C. L. Marley*

The Breed That Didn't Make It

THE AMERICAN CREAM was a breeder's dream.

Unfortunately, it pretty much stayed a dream. And it never really developed into the meaningful draft horse breed some folks had hoped for.

While the American Cream horse breed is practically non-existent today, a few old-time draft horse fanciers still know where you might find one or two of the rich cream-colored horses.

Early-day breeders pointed with pride to the fact that the American Cream was the only draft horse breed ever developed

> "The American Cream was the only draft horse breed ever developed in the United States..."

in the United States. But over the years, the breed wasn't to be.

Goes Back to Early 1900s

The forerunner of the breed was a draft type mare of unknown ancestry. Nobody remembers much about her except for her outstanding rich cream color. This mare lived on a central Iowa farm during the early part of the twentieth century.

As her offspring foaled and grew up, farmers soon realized the old mare continually bred true to her cream color. In fact, breeders found the rich cream color was maintained even when her offspring were mated to other draft breeds to improve type and quality.

Yet it wasn't until around 1935 that any special efforts were made to develop the cream-colored animals into a distinct new draft horse breed. Around that time, a few farmers began linebreeding and inbreeding the cream horses with hopes of establishing a brand new draft breed, a first for the United States.

The years of breeding work paid off when the National Stallion Enrollment Board recognized the American Cream as a new draft horse breed in 1948. Two years later, the breed was recognized by the Iowa Department of Agriculture.

Organized in 1944

Interested breeders met in the spring of 1944 at Iowa Falls, Iowa, to organize the American Cream Horse Association. Several months later, the state of Iowa approved the group's charter and an association office opened for registering and transferring business.

Yet there was not exactly a deluge of animals to be registered. During its first 16 years of operation, in fact, the association registered only 196 horses. In 1960, only two horses were registered. And no American Cream

> "Pink skin was the key to producing the rich cream color..."

horses have been registered since 1972.

The ideal American Cream had a white mane, white tail and was medium cream-colored. It also had pink skin and amber colored eyes.

The pink skin was the determining factor in producing the rich cream color. Breeder experiences indicated dark-skinned American Cream horses frequently would not produce a satisfactory cream color.

From White to Amber

Another unusual and distinctive trait was the amber-colored

Frank Lessiter

THE AMERICAN CREAM draft horse was always recognized by its rich cream coloring. The breed started from a cream-colored mare whose offspring always carried the cream color. Unfortunately, very few of these horses are still around.

eye. Interestingly enough, colts were born with nearly all-white eyes. Yet these eyes would begin to darken in a short time. And by the time the horse reached maturity, the eyes would have turned amber in color.

The American Cream was a medium-heavy draft horse. Average weight for mares ran around 1,600 to 1,800 pounds. Stallions weighed 1,800 to 2,000 pounds.

Breed Didn't Make It

Many people in recent years have wondered why the American Cream breed failed. One of the reasons was that the new breed originated in the late 1940s and early 1950s, just when

---

"During the group's first 16 years, only 196 of the American Cream horses were registered..."

---

draft horse numbers declined sharply. And there weren't enough American Cream horses to keep the breed going when the bottom fell out of the draft horse business.

"The demise of the breed is a strange and discouraging situation," says Karen Topp, a central Iowa gal who served as association secretary. "They were nice horses and of good quality. They were practical since they weren't huge. And they were beautiful with that rich cream color and good disposition.

"But they were just a bit too late for the times. They simply came along too late."

Lead a Horse to Water...

"YOU CAN LEAD a horse to water, but you can't make him drink." That's a popular saying, yet it's not the right quotation. The original, penned many years ago by John Heywood, goes like this: *"A man may bring a horse to water, but he can't make him drink unless the horse wants to."*

Most horsemen always felt it was best to water a horse frequently. Yet it could be dangerous to give a horse a big swig of water if he was hot or hadn't had a drink for some time. Teamsters often argued about whether to water a horse before or after feeding. Sometimes it seemed best to just put out the feed and water, then let the horse decide.

A TYPICAL HORSE puts away 10 to 12 gallons of water a day, say the experts. But many horses seem to drink twice as much water.

J.C. Allen and Son

MANY A watering tank was built so horses and other farm animals could drink from both sides of a fence. Some community troughs were built in corners where animals in up to four pens could drink from a single tank.

GASOLINE ENGINES powered some early-day water pumps. This certainly took the backache out of pumping water by hand for a number of thirsty horses.

J.C. Allen and Son

J.C. Allen and Son

J.C. Allen and Son

IT'S A BIG watering trough for just one team of Percherons. Built from stave silo block, this cattle watering trough held hundreds of gallons of fresh water.

92

A HOT SUMMER day often meant a farm boy could cool off his bare feet in the watering trough while his brother watered the horses that had worked hard all morning in the field.

J.C. Allen and Son

Frank Lessiter

OLD BATHTUBS that had seen better days in the farm house often ended up their career out in the barnyard. Here, they took on new life as fancy porcelain horse watering troughs.

PERCHERONS ARE the preferred breed for the Bob Robinson family. Bob, who works full-time for an electrical contracting firm, and his wife, Marilyn, keep around 20 Percheron horses on their 270-acre farm.

BELGIANS CAPTURED the heart of the Bob Dunton family. Bob, a United States Department of Agriculture worker, his wife and two sons have found great satisfaction in raising big horses.

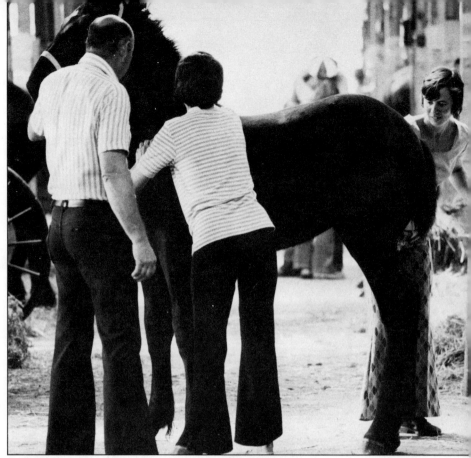

Frank Lessiter

Ed Schneckloth

The Return to Real Horsepower

CLYDESDALES WERE picked as the best breed by the Richard Wegner family. Wegner, a school superintendent, shares the work with his wife and daughter, Carla.

Frank Lessiter

WHAT DO A school superintendent, an electrical contractor manager and a federal government worker have in common?

Raising horses? Yes.

But if you guessed these three men simply raise pleasure horses for riding, you're dead wrong. Instead, these Michigan mini-farmers have a common love for breeding and showing the big horses that farmers used decades ago to work their fields.

If it weren't for draft horse enthusiasts like these, the mammoth animals today could well be nearly extinct.

In 1918, 26.7 million draft horses were used to pull the machinery that worked American farms. Today, they've been replaced by big tractors that busily march across our nation's fields, doing the farmer's work faster and more efficiently. And the number of draft horses and mules has shrunk to under three million head.

Yet, interest in breeding and showing draft horses is again on the upswing. And it's the folks getting into the draft horse business as a hobby who are the ones

---
"When you raise draft horses as a hobby, there's no limit to what you'll pay for a horse..."
---

giving the heavy horse much of its momentum these days.

An After School Hobby

Richard Wegner of Clinton, Michigan, has been raising Clydesdale draft horses for only a few years. This newcomer to the draft horse business is a school superintendent.

"Coming home and working with the draft horses is a complete change from the school superintendent's job," he says. "It's hard work, but after a day at school, physical work like this may well be the way to keep in shape and avoid a heart attack."

Wegner remembers growing up on a 120-acre farm when draft horses were still king. But then

LAUNDRY WORK takes a great deal of time on the white feathered feet of the Clydesdales. Carla Wegner spends hours washing and grooming the beautiful white feather on the legs for the next show ring performance.

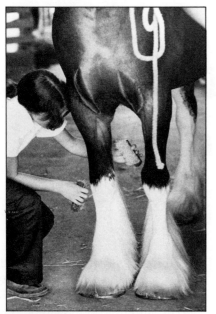
Frank Lessiter

Frank Lessiter

Frank Lessiter

AFTER A DAY behind his school desk, Richard Wegner finds the hard physical work that goes with the horses is good for his health. The horses help keep him in top-notch shape while he enjoys the thrill of victory in the show ring.

he went away to service, college and a teaching job.

He lived in town until 9 years ago, when he got the itch to return to the country. He bought an 85-acre farm and now rents the cropland to an area farmer. But he uses the barns to house his Clydesdale horses.

The horse hobby didn't begin

CONCENTRATING MAINLY on halter classes with the family's Clydesdales, Richard Wegner finds draft horses a good hobby for the whole family.

"My wife enjoys showing the horses, too. You soon become good friends with other draft horse owners. She does the cooking and keeps our horse trailer in order when we are on the road. And she keeps the tack room organized at home."

Family's Greatest Thrill

Wegner says the biggest thrill for the entire family comes at foaling time. "We take turns going out every hour to check on the mares day and night," he says. "It is really a thrilling time when you see a little colt being born."

Wegner firmly recommends the draft horse breeding business to others. "Somehow or other it will enrich your life," he says. He admits it will mean plenty of hard work. But if you have a job where manual labor is not one of the requirements, then that hard work and what you achieve with it are part of the satisfaction.

"You can get plenty of help in getting started. There are a lot of nice people in the draft horse business. And the longer they have been in the business, the more they seem willing to help newcomers learn about draft horses," he says.

The enthusiasm apparent in Wegner's face as he says that makes his next comment almost unnecessary: "We really enjoy the draft horse breeding business. It's a good clean hobby that the whole family can get involved in."

Nuts to Profit

As partial as Wegner is to Clydesdales, Bob Dunton of Saranac, Michigan, is to Belgians. He has been breeding Belgian draft horses for nearly 20 years. An employee of the Federal government's Agricultural Stabilization and Conservation Service office

until Wegner and his wife bought an American Saddlebred horse for their daughter when she was 10 years old.

"We wanted her to take English riding lessons and really learn to ride well," he recalls. "We soon found she liked horses, but wasn't really interested in learning to ride.

Took $1,500 to Start

"Then four or five years ago, I got the urge to buy some draft horses. I don't know why . . . I just did. I had always felt the Clydesdale was the best breed.

"My daughter had helped her uncle show some Belgian horses at some of the fairs. She saw the Clydesdales at these fairs and liked them, too. So we talked it over and plunged in."

First, Wegner bought a $1,500 mare to get started. Today, with draft horses making a comeback, you couldn't buy one like her for less than $3,000.

Working with the horses is definitely a family affair for the Wegners. It generally takes about an hour in the morning and another hour at night to feed and care for the horses. Wegner also tries to hitch and actually work the Clydesdales each weekend.

He schedules his vacation time so he and his family can show their horses at four fairs each

---

"Folks getting into the draft horse business as a hobby are giving the heavy horse much of its momentum . . ."

---

summer. In addition, they try to attend several more draft horse shows each year as spectators.

During show time, his daughter, Carla, and Wegner split the work. "Carla is excellent at washing the white feet and rolling the manes for showing," he says. "I groom the horses. Both of us show them.

Frank Lessiter

A FEW CENTS per hour is all Bob Dunton figures his family earns from draft horses after adding in all labor costs.

Edward Druck

at nearby Ionia, Dunton keeps 20 to 25 Belgian draft horses on his 62-acre farm.

The entire farm is now seeded to pasture and hay. Dunton used to raise some wheat, but now concentrates on raising strictly forage for the horses.

The horses are a family affair for the Duntons, too. Bob's wife, Mary Jean, is a registered nurse, so she's able to give the horses all of the necessary shots and does most of the minor veterinary work. Their two teenage sons, Richard and Randall, do much of the showing of the horses.

"I've been interested in draft horses all of my life," says Dunton. "I was born with them, having grown up on a farm.

"But when I got out of the army, I went to college and then hired on with the United States Department of Agriculture. It wasn't until I was transferred here in 1958 that I again got interested in breeding draft horses."

Dunton says the horses are raised as a hobby—certainly not for income. "We show a little profit, I guess, from showing the horses and from selling some breeding stock. But we are probably only earning a few cents per hour for our time if we were to figure in all of our labor costs.

"It's something you do mostly

HAVING A NURSE in the family helps the Bob Duntons hold down health costs with the horses. Bob's wife Mary Jean, a registered nurse, gives most of the needed shots and handles most of the minor veterinary work. But everyone pitches in when there is horse work to be done.

PASTURE AND HAY are the only crops grown on the Bob Dunton family's 62-acre farm. These forage crops are fed to the farm's 20 to 25 Belgians.

J.C. Allen and Son

Frank Lessiter

THE BELGIANS are trucked to a number of shows each summer. The family's two sons, Richard and Randall, help show the horses at most shows.

as a hobby and for pleasure rather than the money."

Some Problems, Too

While there are plenty of advantages to raising draft horses, Dunton is honest enough to cite some disadvantages.

"The horses have to be fed twice a day if they are kept in the barn," he says. There is also a sizable quantity of manure that has to be disposed of in one way or another.

"Plus, these big horses have big appetites. Some of these horses can stow away a bale of hay a day—and hay isn't cheap these days.

"I guess there is also always a certain risk of injury in handling large animals. I know it certainly hurts when a 2,000-pound horse steps on your toe."

Part of the pleasure the Dunton family gets out of raising draft horses is the opportunity to compare their horses and talk with other breeders at various state and county fairs. The family usually exhibits their Belgian horses at eight shows each summer.

Big Horses a Hobby

Are Clydesdales or Belgians best when it comes to raising draft horses? Not if you talk to

BOB ROBINSON grew up with Belgians. But he switched to Percherons after marrying into a family that has raised them since 1918. All 20 of the family's horses trace back to a mare his wife had as a 4-H Club project.

Bob and Marilyn Robinson of Richland, Michigan. They raise Percheron horses and show them at nine shows a year.

Bob runs the Millright Division for Rowen and Blair, an electrical contractor in Kalamazoo. They live on a 270-acre farm and raise hay, corn and oats. But farming's still a part-time occupation—their income chiefly comes from Bob's off-the-farm job.

Both Bob and Marilyn grew up in the draft horse business. Bob's family raised Belgian horses. But

> "A few hours of work with the draft horses may keep me from having a heart attack..."

after getting married, Bob switched to his wife's preference, Percherons. Marilyn's father had started raising purebred Percheron horses in 1918.

"Marilyn started with a Percheron mare in 4-H Club work when she was 12 years old," explains Robinson. "All 20 Percheron horses we have today are direct descendants of that mare."

During the show season, Marilyn and the horses are usually away from home for nine weeks. Bob commutes back and forth to the fairs from his job—usually hitching a ride with other horse breeders or flying back and forth with fellow members of the local flying club. He shows the horses.

The horses are hauled with a pickup truck towing a 20-ft. gooseneck trailer which can hold nine horses. They also take a large motor home with them to stay in at the various fairs. "When you're on the road for nine solid weeks like my wife is,

Frank Lessiter

you *need* a large motor home—you don't want to be cramped," explains Robinson.

Gets in Your Blood

"The showing is exciting," says Robinson. "It gets in your blood just like anything else. It is definitely a hobby. You can just about break even on the premium money paid at various shows around the country."

Besides exhibiting the Percherons in the various halter breeding classes, Robinson also drives a mare in a cart class.

"We take the wheels off the cart and it fits right in the front of the gooseneck trailer," he says. "We usually borrow a harness from another breeder, so we

don't have to lug that with us."

Robinson's farm is laid out so the horses pretty much take care of themselves. They generally put a stallion and a few mares in the barn during the winter. But the remainder of the horses stay outside the year around.

"We've got horses that haven't been in the barn for 10 years." says Robinson. "They run on 100 acres of pasture that has a pond so they can water themselves.

"We run them on corn stalks during the winter and this provides plenty of cheap, but good, feed. They really take care of themselves."

Robinson has a good explanation as to why draft horse prices are on the increase. "When farmers worked horses in the old days, you could only afford to pay just so much for a good horse," says Robinson. "But when you are raising draft horses and really love this hobby, I guess there's no limit to what you'll pay for a horse."

H. Armstrong Roberts

SNOWY WEATHER didn't mean the horses got to rest. Sleighs replaced wagons as the horses continued their year-around work. There was wood and manure to haul down the bright snow-covered farm lanes.

AN AMISH farmer rotary hoes his corn crop. Many Amish fa[rmers] still rely on "real horsepower" to handle their field work.

Dale Stierman

J.C. Allen and Son

When Horses Tilled the Land

EARLY DAYS on the farm meant plenty of hard work in the fields for both farmers and draft horses. The hours were long and the days were often hot. But you had to keep "pushing" to get all the crop work done on time.

Plowing, tilling, planting, cultivating and harvesting often required the longest days in the field...stretching all the way from sunup to sundown. Such a dependency on horsepower usually brought a farmer very close to his team—he knew his team well, and the team likewise knew him.

Many a grown man today can probably recall how as a youngster he helped harness or unharness his dad's team in the black of early morning or late evening. During the busy planting and harvesting seasons, this harnessing and other chores were done by lantern light. *Every possible*

PLOWING OUT potatoes in the fall was often a family affair. Dad handled the touchy plowing job while the oldest son drove the team. Typically, the job of picking up spuds fell to junior.

HARNESSING AND HITCHING a green horse was never easy. These farmers teamed a 10-year-old gelding with a two-year-old filly for training.

J.C. Allen and Son

State Historical Society of Wisconsin, John Murray for Milwaukee Journal

H. Armstrong Roberts

hour of light was spent in the field during the critical farming periods.

Even during the winter months when the farmer got a rest from crop work, he and his team simply turned to other waiting tasks. There were tons of manure to be hauled—often on days when temperatures were well below the freezing mark.

Horse Held Its Own

When tractors appeared on the scene, the horse still held its own for many years. Horses continued to compete favorably with tractors . . . up until farm labor got scarce during World War II and the race to more mechanization on the American farm began in earnest.

Admittedly, though, there had been a sharp drop-off much earlier in draft animals. The number of draft horses and mules on American farms dropped by

HORSES HELD THEIR own against tractors until World War II brought serious farm labor shortages. Then the race to mechanization began in earnest.

MULTIPLE hitches let farmers plow more ground faster. This 12-horse hitch pulled a four-bottom plow at a fast clip once the fall harvest was completed.

J.C. Allen and Son

H. Armstrong Roberts

J.C. Allen and Son

COMING HOME at the end of a long day, this farmer drove his Percherons across the yard toward the barn.

SPRING PLOWING of this old corn field took place in the Blue Ridge mountains of Virginia. A good man with one horse could plow one acre per day.

ALFALFA SEEDING TIME rolled around in late summer on many a farm. Working a finely-tilled seedbed for the last time with a disc, farmers would hope for a quick shower to bring the alfalfa sprouting up out of the ground within a few days.

HORSES OR TRACTORS? The contest to see which form of horsepower could best handle the large variety of farm jobs went on in earnest during the late 1930s and the early 1940s. The horse seemed to be ahead in the corn cultivating match shown here. But the tractor would eventually be declared the ultimate victor in the years to come.

J.C. Allen and Son

Bob Taylor

more than five million head from 1910 to 1927. Horse prices also dropped sharply from an average of $146 per head in 1910 to only $77 per head in 1927.

Farming Was Growing

The mid-1920s was a time in our history when younger farmers were itching to find ways to boost output and efficiency. They wanted to *expand*, yet couldn't without finding new ways to turn out more field work in a day's time.

These young farmers weren't as tradition-prone—they weren't satisfied with doing farm work the same old way. Instead, they wanted to finish farm jobs as quickly and cheaply as possible.

Using the "field experience" and imagination farmers have always been known for, they soon found that by hitching more than two horses, they could do more work in a day's time. Instead of turning just one furrow with a two-horse plow, they wanted to turn two, three or more furrows in one trip across a field with multiple hitches of four, six, eight or more horses.

While bigger hitches such as these came into use in a growing number of states in the late 1920s, it wasn't really a fresh idea. Horse-drawn combines—some with 30 or 40 horses controlled by one driver—had been used in the great wheat-producing areas of the West.

Farmers knew from experi-

106

A FEW MOMENTS of rest were enjoyed by the team while the farmer stopped at the end of the field to refill the drill. Spring was the year's busy season for both men and horses as the working of the land and the planting of the crops brought some mighty long days.

Elma and Willard Waltner

SOME RODE while other farmers preferred to walk along behind the grain drill. Note the toolbox nailed to the rear platform and the wooden wheels that made many earlier-day drills go.

J.C. Allen and Son

HAY MOWING went fast on this Nebraska farm with three teams, three mowers and three men all working in the same field to pull down the alfalfa.

HAY LOADING could be hot, hard work. It took two men to for the hay into a load as the forage came up the hay loader while third—often a young farm boy—kept the team straddling the windrow

J.C. Allen and Son

MOM WAS a part of many haymaking crews. A sunbonnet helped this lady of the farm keep the hot sun and hay leaves off her head as she handled the team on the dump rake.

BUCKRAKING was the way much of the hay was handled in western ranching areas. A number of 8 to 10 foot long wooden tines stretched out on the ground between the horses swept the hay into large piles as the team moved ahead. The hay would later be tossed into large stacks scattered out on the prairies of these ranches.

Bob Taylor

J.C. Allen and Son

J.C. Allen and Son

J.C. Allen and Son

J.C. Allen and Son

MARKETING STRAW was a big business for these Indiana farmers. The permanent seats and roofs found on these six wagons kept the chaff and dust out of the eyes and hair of these farmers. Yet they were a luxury many farmers never enjoyed on their wagons.

THROUGH A SERIES of ropes, pulleys and steel tracks along the peak of the barn, huge forkfuls of hay could be lifted into the big barns and dumped in the right mow by the power of a team of horses. Setting the hay forks three or four times meant a whole wagon load of hay could be moved to the mow.

109

State Historical Society of Wisconsin

Dale Stierman

J.C. Allen and Son

BUNDLE HAULING TIME meant moving the grain to the farmstead for stacking. But there was always a moment to enjoy a refreshing drink of cold water in the field.

ence they could plow two acres of ground a day with a walking plow and a two-horse team. But they also knew they could do much better in a day's time with bigger hitches.

Using an eight-horse hitch, a farmer could plow 8½ acres, disc 40 acres or harrow 80 acres in a day's time. Energetic young farmers marvelled at the excitement of such output.

Soon more farmers were using the bigger hitches to reduce crop production costs, mostly by saving valuable time which let them plant and harvest more acres, or finish their existing acres much earlier.

Multiple-horse hitches also kept the more expensive tractor from moving quickly into the farm scene. Hay was far cheaper than gasoline. So multiple hitches allowed farmers to retain the reliable four-legged form of drawbar horsepower which was

110

165 HORSES pulled these five Holt grain harvesters in the 1890s on the Drumheller Farm near Walla Walla, Washington. These huge combines, with the driver seated out over the teams, were used mainly in the northwest where the wheat crop dried down at a rapid pace.

CUTTING GRAIN with the twine-tieing binder that remained remarkably unchanged for more than 50 years.

J.C. Allen and Son

Milwaukee Public Museum

THRESHING TOOK lots of manpower and plenty of horsepower. This led to formation of neighborhood threshing rings which often used 15 men and two dozen horses for hauling bundles to the thresher and grain to town.

SILO FILLING could be a nervous time for horses—especially if a stone went through the machine.

HAULING GRAIN to town to be ground into feed was a weekly ritual on many farm The wagon would be piled high with sacks of ground feed on the return tri

Stewart Doty

State Historical Society of Wisconsin

DRYING TOBACCO was a two-part job. The tobacco was first allowed to dry in the field. Next, the tobacco was hauled to a tobacco barn for still more drying before being sold.

J.C. Allen and Son

HAULING HOGS to market was done with high wagons equipped with livestock racks back in 1913. Junior takes a last look at the farm's finished hogs as Dad prepares for the 15-mile trip to town.

J.C. Allen and Son

A FAMILY PORTRAIT? Perhaps as this farm family posed. Many teamsters were as proud of a team as of their kids.

J.C. Allen and Son

Milwaukee Public Museum

SUNDAY BEST was the look as this farm family stopped on their way home from church to have their photo made out in the pasture among the cows.

Dale Stierman

AMISH FARMERS still rely on draft horses to farm their land. They refuse to give in to the virtues of mechanical horsepower.

Edward Druck

SKIDDING LOGS is a task draft horses excel at. Oftentimes, they can work in wooded areas where tractors simply can't maneuver.

self-replacing, consumed homegrown feed instead of gas and had maximum flexibility.

Easy Way to Hitch

Driving a multiple horse hitch was not as difficult as it appeared. In fact, some old-timers say that with a little practice the horses could practically drive themselves in the field. There was practically automatic control of the wheel team, since this team was "tied in" and "bucked back" to the other horses

Light-link halter chains with simple snaps on the ends tied the wheel team to the traces of the team ahead. Ropes or straps were

Dale Stierman

J.C. Allen and Son

BRINGING HOME a load of feed, these horses swing into the farm lane at a fast pace. Yet there's no doubt but what the Amish farmer is still in complete control of the whole situation.

SLOPPING THE HOGS was a daily chore. This farmer apparently fed a substantial number of hogs as indicated by the four big slop barrels pulled by his team of Percherons.

115

Northwest Unit Farm Magazines

CUTTING ICE was a winter-time task on many farms. Horses and sleighs hauled the ice to sheds where it would be packed in sawdust for summer.

H. Armstrong Roberts

H. Armstrong Roberts

IT'S HARD to beat a team of horses when it comes to feeding cattle in the winter. A team will drive itself as one or two men toss bales of hay to the hungry cattle. Try feeding hay out on the range with a tractor and wagon and you'll immediately have to add a driver to your crew. Besides, a team and sleigh can work in deep snow without getting stuck.

KEEPING THE fires going was another area where horses really shone. Horses were unsurpassed when it came to hauling firewood out of the woods with a sleigh.

116

MAPLE SYRUP season brought out farmers on snowshoes and horses hitched to sleighs to gather the sap on late winter days. Horses wandered through the trees with ease as farmers quickly dumped the buckets of sap hanging on the maple trees.

TIMBER cutting could always be done when there weren't any other farm jobs to be handled during the winter. Whether cutting firewood or timber for a new barn, winter was the best time of year to skid logs. Mud problems were avoided by skidding logs on frozen ground.

H. Armstrong Roberts

H. Armstrong Roberts

used to "buck back" the teams for even more control.

Most multiple-horse hitches were used tandem-style—where four or more teams were hitched one ahead of the others. Any needed adjustments could be easily made simply by taking up or letting slack out of the eveners. The slowest team was usually hitched as the wheel team and the quickest horses were hitched out front.

Many farmers handled most of their field work with four, six or eight horse hitches during the late 1920s and the 1930s. These multiple hitches helped farmers improve their efficiency during the depression years, when such efficiency was badly needed.

But the multiple hitches fell out of favor as the 1940s rolled around . . . and tractor power started replacing more and more real horsepower in the fields.

Stewart Doty

LATE NIGHT cleanup was often a common task during the long sunup to sundown spring planting season. Needed light came from a kerosene lantern that slid along a wire. The work slackened off a bit by winter, but there was still plenty of manure to fork.

State Historical Society of Wisconsin

Plowing with Horses For the Fun of It

WHEN MOST FOLKS think of plowing with horses, they visualize a single team hitched to a walking plow.

But horse plowing for Clarence Nordstrom means harnessing up to eight horses. This Pine City, Minnesota, farmer covers a good bit of ground with eight horses hitched to a three-bottom gang plow—he plows as much as eight acres in a day's time.

Most horsemen would tackle a three-bottom gang plow with seven horses. Yet Nordstrom prefers to use eight horses for several distinct reasons.

"Most other horsemen use four horses right next to the plow with three horses out front," says Nordstrom. "But I like to use three rows of horses—three horses in the back next to the plow, three horses in the middle and two horses out front.

"The eight-horse hitch keeps your line of draft more in line with the center of your plow. You don't get side draft this way."

Another reason for using eight horses instead of seven horses on a three-bottom gang plow, Nordstrom says, is that you need to do things differently to catch the eye of the judges in today's plowing contests.

Nordstrom, a seasoned pro at horse plowing, enters all four classes in the Minnesota statewide contest each year. This means plowing with a walking plow, a sulky plow (a one-bottom plow with the plowman riding a seat on the plow instead of walking), two-bottom gang plow and a three-bottom gang plow.

Nordstrom farms 240 acres with Belgian horses. He uses the

NEITHER RAIN or snow could stop a plowing contest. This farmer kept going with his four bottom plow and 12 horses despite the rainstorm.

J.C. Allen and Son

horses to do all of his plowing, discing, planting and haying.

Interest in plowing with horses is again gaining popularity. "Many of us horsemen figured it was about time to show the practical value of the draft horse," explains Nordstrom.

So they started the Minnesota State Plowing Contest in 1965. "We've got a good thing going now and it has really helped stir up a tremendous amount of renewed interest in draft horses," explains Nordstrom.

Some 60 contestants plowed 30 acres of ground during one of the recent Minnesota state contests. "We figured we should show draft horses working in the fields instead of just exhibiting

---

"We do everything with horses except combining and baling..."

---

them at shows and putting them in parades," says Nordstrom.

"Horse plowing shows their usefulness as a work animal. That's the way this country was built... and field work can still be done this way."

Regardless of what size plow is used, a good rule of thumb is that a man can plow one acre per day for each horse he is using. A team on a walking plow could do about two acres per day. Or Nordstrom could plow eight acres per day with his eight horses hitched on a three-bottom gang plow.

By comparison, a 300 horsepower four-wheel drive tractor

PLOWMEN HOLD their lines in many different ways. Some string the lines through the thumb and forefinger of each hand while gripping the wooden walking plow handles. Yet other plowmen, such as this farmer in an Ontario plowing contest, prefer to wrap the lines around the waist and knot them.

Frank Lessiter

"I'LL BET WE ARE DOING BETTER" seems to be on the minds of these horses as they watch competitors turn the soil. Both horse-drawn rigs seem to have scared the tractor-powered plow off to a safe distance.

"OUT OF THE WAY or we will plow you down" is the message this horse seems to be giving this plowing contest spectator. Working on a backfurrow, the team plunges ahead with reckless abandon as they push the plow up to full speed.

Frank Lessiter

Frank Lessiter

pulling 12 bottom plows can turn over better than 110 acres of land in a long day's time. But a man and horses are limited to 8 to 10 hours work per day.

Experience Really Counts

A man's plowing skill—both in the old days and today—was judged by how well he could open a field with a near-perfect back furrow and how well he could follow with an unending ribbon of straight furrows.

To do a good job of horse plowing takes plenty of talent and considerable practice. "It takes experience, I'll tell you that much," says Nordstrom.

"And it also takes... well, who knows exactly what it takes? There are so many things

═══════════════════════════════
"A horse can plow one acre per day whether on a one, two or three-bottom plow..."
═══════════════════════════════

that go into plowing that you just can't pinpoint all of them.

"I'll say this, it takes a desire to do a good job. You've got to have a good plow and you've got to have horses that know how to handle your particular plow.

"The teamster is also very important. A good horseman can get along with poor horses. But the poor horseman has got to have the best horses there are."

Scorecard Makes Winners

Some 70% of the total score in a plowing contest is usually based on the plowing itself—correct depth, trash covering, firmness of the plowed ground, final finish, how the plow hooks in and out of the headlands and a half-dozen other measures of good plowing.

The other 30% of the score is

based on the appearance of the horses, how they handle the plow and how they are groomed. While many plowmen use extra fancy harness, this isn't considered in scoring the horse and plowman. But there are often

---

"The poor horseman has got to have the best horses there are to plow..."

---

special awards for horsemanship and best-matched teams in plowing contests.

Plowing contests operate in different ways. Each contestant plows his own piece of land in some contests. Yet teams follow each other over the same land in other plowing contests. While this type of contest is harder to judge, spectators can easily see each team plow without having to run from one end of the field to the other.

IS IT STRAIGHT or not? Every farmer always wanted to be known around his neighborhood for the straight furrows he left in his field. And when it came to entering a plowing contest, he wanted to leave even straighter ribbons of soil as he moved slowly across the field.

Frank Lessiter

Stewart Doty

FANCY HARNESS isn't supposed to earn you any extra points in a plowing contest. But it sure looks nice and helps attract the attention of both the contest judges and the spectators. And in some contests, special prizes are awarded to the most attractive teams.

WAITING HIS CHANCE, a plowman rests his horses and studies the plowing ability of a fellow competitor starting a back furrow in a new field.

Frank Lessiter

125

Plows Have Long History

MORE THAN 375 patents had been assigned to plow changes and improvements by 1855. Wooden plows were used until the cast-iron plow came in 1808. John Deere's steel plow was developed in 1837.

THE HORSE-DRAWN PLOWS—mostly 14 inch wide bottoms—used today in plowing contests are authentic implements used for many years to break farm ground.

The three-bottom gang plows carry names like John Deere and International Harvester...plus Rock Island, Emerson and six or seven other mostly forgotten names of plow builders.

"It's obvious these plows aren't something people have

J.C. Allen and Son

FOR MANY FARMERS, most plowing was done with a team and a one-bottom walking plow or a one-bottom sulky plow. Some mighty steep hills were worked with these plows.

THREE AND FOUR bottom gang plows never were used much by eastern farmers. So farmers from this area really enjoyed seeing these big plows and multiple horse hitches competing in plowing contests.

J.C. Allen and Son

H. Armstrong Roberts

thrown together just for the sake of having a three-bottom plow for contest plowing today," says Clarence Nordstrom of Pine City, Minnesota. "I've owned about a half-dozen three-bottom gang plows over the years.

"Many folks from the East or South have never seen a three-bottom gang plow. They farmed earlier in their areas mostly with a horse-drawn walking plow or a one-bottom sulky plow. So seeing these horse-drawn gang plows is of real interest to these people."

Most plows have remained in good condition despite many years of use. Plowshares and moldboards wear out. But the horse-drawn plows themselves keep right on turning over the earth.

Wooden plows were used in the early days of this country. Then came the horse-drawn cast iron plow, patented in 1808.

Yet many people fought the cast-iron plow—believing it

> "Many people fought the cast-iron plow—believing it poisoned the land, caused weeds and drained strength from the soil . . ."

"poisoned" the land, caused rapid weed growth and drained the strength from the soil.

The cast-iron plow was fine for turning the virgin soil of the prairies. But in following years, the soil would stick to the cast-iron moldboard. The sticking soil would have to be scraped by hand from the plow after every few feet of plowing.

So, a better plow was needed . . . one with a smooth moldboard that would prevent the sticking soil problem. The solution was John Deere's steel plow, formed by the Detour, Illinois, blacksmith from a steel sawmill blade in 1837.

After Deere's steel plow came on the market, plow sales really soared. By 1855, over 375 patents had been assigned to plow changes and improvements.

DESPITE MANY YEARS of hard use, most horse-drawn plows have remained in good working order. The plowshares and moldboards wear out, but the plows keep right on going.

J.C. Allen and Son

WRESTLE ONE of these plows for a few hours and you'll want to enjoy a minute or two of rest at the end of the field. And the horses enjoyed—and needed—the rest just as much as did the teamster. Perhaps the farmer was thinking of brighter days ahead—when the crops would be safely in the ground and work days would be shorter.

WHEN A SIX-HORSE hitch was only pulling a two-bottom plow, you knew the farmer was plowing in heavy soil. You could also tell it was a cold fall day by the warm dress of the farmer.

State Historical Society of Wisconsin

J.C. Allen and Son

J.C. Allen and Son

HANDLING TWO TILLAGE JOBS at once was the idea being tried out by this farmer. He had the needed horsepower to pull a trailing harrow behind the plow with eight horses. With two such rigs, he could plow 14 to 15 acres a day.

Get a Horse

J.C. Allen and Son

MUDDY FIELDS AND ROADS often meant many of the new-fangled tractors and autos had to depend on a reliable team of horses to get back to firm footing. There's the old story about the motorist who pays a passing farmer $10 to pull his car out of the mud. "At these prices you must be pulling people out of the mud day and night," the motorist tells the farmer after his car is back on dry ground. "Can't," says the farmer. "At night, I have to haul water to fill the holes in the road."

H. Armstrong Roberts

The Unmatched Excitement of Horse Pulling Contests

SNORTING, puffing and sweating—that's the kind of real horsepower you'll see in any horse pulling contest.

Take in a horse pull and you'll quickly see the horses are eager to pull. They are *competitors* from the word "Go!"—as their powerful legs dig down deep into the dirt in choppy short strides as the weight moves forward... and keep pawing ahead until they pull the load the required distance or the driver realizes the team can't handle the load and pulls them to a stop.

Horse pulling is like a fever with some horsemen and with some spectators, too. You'll likely find a big crowd anytime teams start harnessing up for some pulling action.

It's one of rural America's big sports... one that's making a comeback in many areas. Better

ANOTHER DAY, another horse pull. Some teamsters and their teams make as many as 30 or more horse pulls a year. They compete for the fun of it.

Frank Lessiter

CHAINS, ROPES, harness, stock racks and binder twine kept this team in the truck during a short journey to a horse pull. Other horses, on hand an hour or two earlier, contentedly munched a nice tasty fill of grain.

Frank Lessiter

Frank Lessiter

Frank Lessiter

than 200 horse pullers belong to just one state association alone. They had their pick last summer of 32 horse pulls in a month's time—even eight pulls on the same day within that state!

It's a sport and hobby for these men and their massive horses—some with chests big

---
"Can't touch horses during a pull. Some rules say you can't swear or even shout..."

---

enough to cover any two men. These are big horses, the kind that often stand better than 6 feet tall at the withers.

An Exciting Sport

Few athletic contests can match the excitement of seeing big horses of any breed or color—and many mixtures of both—put their shoulders to the harness as they move tons of weight down the field.

What determines horse pulling success or failure is the team-

work between horses ... plus the rapport between team and driver.

The driver knows exactly how much he can ask of his team—and he uses his voice and his hands on the lines to communicate that knowledge.

Horses that work well as a team can actually sense the moment they are hooked to the weight. With good teamwork, the initial lunge is strong and smooth.

The key item in a horse pull is 27½ feet, the distance a load must be pulled before moving on to the next heavier weight. The 27½ feet distance is the amount of space a horse needs to reach its full horsepower. It was determined many years ago that a

HARNESS UP comes the word and the pullers spring to action. The excitement of another horse pull will soon begin as the gentle giants compete for the ribbons and prize money.

134

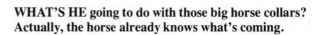

WHAT'S HE going to do with those big horse collars? Actually, the horse already knows what's coming.

LAST MINUTE REPAIRS make blacksmiths out of many pullers. When a shoe comes loose, it has to be fixed on the spot.

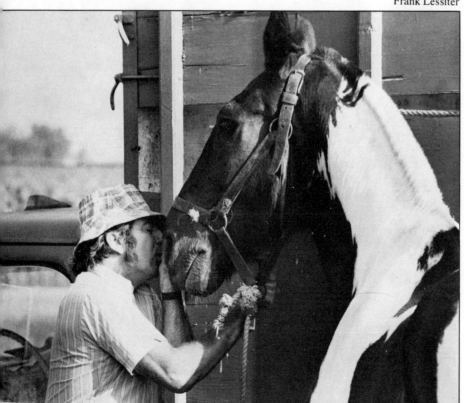

BLOWING A home-made white powdered concoction into the horse's nostrils, this teamster is apparently looking for some extra power or stamina in the horse pull that's about to begin.

horse starts running out of power under a full load after 27½ feet.

Pulling is done with various types of weights on a stoneboard or with a dynamometer. Usually truck-mounted, the dynamometer uses an oil pump and gauge where oil pressure can be increased to boost the weight being pulled.

Not Really Complicated

The rules for horse pulling are quite simple. If a team weighs in at under 3,200 pounds, it is usually considered a middleweight team. Any team that

John White   John White

Frank Lessiter

A QUICK PUFF helps quiet the nerves of Kansas puller Orville Medearis as he watches a pull.

IT TAKES TWO, sometimes three men to get a team into pulling position.

weighs over that figure moves into the heavyweight class. Some teams will weigh 4,600 to 4,700 pounds.

A few pulls feature a lightweight class—usually 2,800 to 3,200 pounds. And a few smaller pulls offer only one class regardless of the weight.

If a team of supposedly lightweight horses checks in a few pounds over the weight limit, a couple of quick jaunts up and down the road may help them squeak in under the weight limit.

Teams usually start pulling with a fairly light weight. The

136

WAITING THEIR TURN to pull, several teamsters exchange friendly comments on which team is going to win. Then a loud "giddap" and it's off to the stoneboat. The chore of getting the excited team backed up to the stoneboat and properly hitched for the pull has begun.

Frank Lessiter

Stewart Doty

"I PULL because I like horses", says teamster Frank Kurth. "But I don't travel far since I've got chores at home."

137

weight is increased until only one team is able to pull it. Each team gets three tries to pull any given weight.

With two or more teams pulling the same weight in the last go-around, the winner may be decided by the team that can pull the weight the farthest—even though neither team makes 27½ feet.

One team might pull the weight 15 feet, while the other team may only make 12 feet. In this case, the team with the longest pull would win.

During the final pull, the number of attempts and misses

> "The cash is small, the trophies nice and the friends great..."

can also be important if none of the teams pull the weight 27½ feet.

Two Horses, Three Men

Besides the driver each team needs a pair of "hookers." While the driver handles the lines and urges the horses on during the pull, the "hookers" are important prior to the pull—they hold the horses back, keep them moving straight and try to have them pulling together before dropping the pins that hook the team to the weight.

It often takes all of a driver's strength to control the eager horses before the pulling actually starts. If the hitch goes smoothly, three things happen practically at once: The driver hops onto the seat, the "hookers" pin the eveners to the load and the horses leap forward with the weight.

Once in a while the "hookers" have an added chore: Chasing a run-away team!

Three Pulls and You're Out

As previously mentioned, each team gets three chances to pull the weight. The weight a team can pull depends on the day, the mood of the horses, the weather and the ground. Most horses do their best on sod.

To conserve a team's strength, a driver may elect to take only one of the three pulls at the easy light weights. By doing this, he can begin the pull anywhere on the track. This lets him choose the area that will give the best footing. But to do this, he must agree in advance to only one pull in this go-around.

If his horses fail to pull that light weight on the single try, he's out of the contest. That's why most drivers opt for all three tries.

If a team fails to pull the distance on the first pull in the regular three-try sequence, the driver can keep the team hitched and immediately try a second

John White

THREE CHANCES and your team is out. The weight a team can pull depends on the day, the mood of the horses, the weather and the ground. Many teamsters say horses do their best on sod.

When a team fails to pull the weight in three attempts, it is out of the competition. Additional weight is added and the other teams continue to pull.

Don't Pull for Money

Men who pull horses do it mainly as a hobby. Many hold jobs in town, yet still like to take the lines in their hands for the exciting thrill of pulling.

While the cash rewards are usually small, the company is good and the trophies beautiful. Most pullers compete for the sheer fun of it—although some first places bring $150 or more—and a few get up to the $1,000

"Studies indicate a horse's strength starts dropping off after pulling 27½ feet..."

---

mark. Some pulls pay money as far down as 25 places, or maybe offer $5 per team for just showing up.

But most horse pullers are lucky to make enough to buy gas for the truck, oats for the horses and a meal or two for themselves.

Getting in Shape

Rugged daily training is the best way to get horses in shape for pulling, according to experienced drivers. The trick is to build up the horses' muscles by steadily increasing the load. But you must avoid "sticking" the horses with a bigger load than they can handle—this tends to discourage them from pulling more weight in the future.

Michael Zibell starts training his horses with a stoneboat and sled. The Stoughton, Wisconsin, puller trains new horses with a

pull after a minute's rest.

If his team fails to make its second pull, the driver can unhitch and place the load anywhere he wants on the field or track for his final pull.

When this happens, you'll see drivers studying the contest area closely—often down on their knees for a closer look at the sod or dirt. They want the area that gives the least resistance for the load and the best footing for the horses.

If an excited team fails to pull the weight on its first attempt, a driver is allowed to unhitch the team to calm them down. But by doing so, he forfeits one of his two remaining chances to pull.

BAGS OF SAND provide the weight on the stoneboat for this early morning horse pull. Starting with an easy-to-pull weight, more bags of sand and large, heavy chunks of concrete will be added until only one team still remains.

EIGHT HANDS work feverishly to get this "raring to pull" team backed up and properly hitched to the stoneboat. Soon three of the men will jump aside as the team quickly lunges into action.

Frank Lessiter

Frank Lessiter

REACHING DOWN for his last energy, one horse trys to handle the weight. Yet the two horses are not pulling together.

"ADD MORE WEIGHT" comes the call from the horse pull judge after several teams pull the previous weight the distance. Eight men skillfully maneuver—slowly to avoid smashed feet or thumbs—a huge chunk of concrete that weighs 700 or more pounds into place.

Frank Lessiter

Frank Lessiter

"HOLD IT, you've made it" and with little room to spare. The team is practically on top of the big infield public address stand.

Frank Lessiter

fairly light load for about four weeks to build muscle tone in the animals.

Next, come a couple of weeks of hitching the horses day after day to larger and larger loads. "You have to build up the horse's confidence," explains Zibell, a third generation puller in his family. "If a horse thinks he can do it, he will."

Zibell and his father, Phillip, make 30 to 40 pulls a year from May through September with their two teams that have enjoyed plenty of competition.

"ALL RIGHT, get going," hollers the teamster as his team gives it all that they've got. Note how he keeps the lines out of the way by tossing them over his right shoulder.

KICKING UP hundreds of clods of dirt and plenty of dust, this team gets its weight into the pull. The teamster stands to the side instead of adding unnecessary weight to the pulling sled.

John White

Frank Lessiter

WOODEN SEATS built on both ends of this stoneboat allow teamsters to pull in either direction without any fuss. There's no need to waste time hauling the stoneboat back to the other end of the track.

According to Zibell, once you get a team trained to pull, you've got to work them constantly between pulls to keep them in shape.

Besides the combination of good horses, tedious training and plenty of patience, horse pulling also requires a steep investment in extra-duty harness. Reinforced hames and double-strength harness straps are needed. Three-inch tugs are essential due to the stress when a team jerks a load into motion.

But it's all worth it once the

143

EAGER TO pull, this team moves out with powerful short, choppy strides in an effort to pull the weight on the stoneboat. Like many old-time teamsters, this farmer has caught "horse pulling fever" over the years.

John White

Lyle Orwig

Frank Lessiter

HORSEPOWER IN MOTION pretty well sums up this horse pulling scene. The slow shutter speed of the camera captured the beautiful power and motion a team puts forth as it pulls the heavy weight down the track.

WITH GOOD TEAMWORK, the initial lunge against the weight is strong but smooth. Experienced horses that work well together as a team sense the exact moment they are hooked to a load.

MOVING INDOORS lets horse pullers keep competing during the cold, winter months. Unlike tractor pulling contests held inside, there's no question with an indoor horse pulling contest of what to do about engine exhaust fumes.

Judith Buck Sisto

ONE OF AMERICA's truly traditional rural sports, horse pulling is making a comeback in many of the country's smaller cities and towns. Besides being a sport, it's the hobby of a growing number of full-time and part-time farmers alike.

John White

Frank Lessiter

SLIGHT CONFUSION crops up at times. This team isn't sure which way they want to go, as indicated by the disgruntled look on the teamster's face.

growing popular sport of horse pulling gets in the blood of both you and your team.

41 Pulls, 20,000 Miles

During a recent summer, horse pull official Steve Lance worked 41 pulls with his dynamometer. Driving better than 20,000 miles, he worked pulls in a six-state area around his Watertown, Wisconsin, home base.

Yet providing the dynamometer for horse pulls is just a part-time job for Lance. He's a buyer for Iowa's Dubuque Packing Company in his regular four-day-a-week job, buying around 45,000 calves yearly at the Milwaukee Stockyards. Since getting into the calf buying business in 1952, Lance has bought better than one million calves in Milwaukee.

The biggest weight ever pulled on Lance's machine is 3,900 pounds... equal to a team pulling 50,600 pounds of wagon weight on a paved street.

Lance feels the dynamometer gives a more uniform pull than a

> "You have to build up the horses' confidence gradually..."

stoneboat. "The dynamometer doesn't give a friction pull like a stoneboat does," he says.

Explaining how a dynamometer works is no easy task. Simply put, it works much like twisting a big rubber band. The more you twist the band, the harder it is to pull.

With a dynamometer, the truck is on one end and the horses on the other. Both try to overcome the resistance built-up by oil pressure in the dynamometer pump.

Asked about some of the worst pulling conditions he has ever seen, Lance remembers one hard, packed track used for hot rod races. "The horses couldn't dig their hoofs in at all to get a

> "A horsepower is equivalent to raising 33,000 pounds one foot high per minute..."

start with a load," he says. "It was awful."

Sod usually provides the best pulling conditions, but is not always available at horse pull sites. And even sod can be "rock hard" during a really dry year.

While dangerous at the time Lance now laughs when he thinks about an incident a few years back during a Wisconsin State Fair pull.

"One driver had a tight hold on the lines as he climbed onto the seat of the dynamometer," recalls Lance. "When the horses started to pull, they jerked him right off the seat onto the whippletress. He stood right there on the eveners, balancing carefully during the whole pull.

"If he had ever fallen off, the truck would have run over him. He was scared. I've never seen him hold the lines tight since then."

HOW MUCH DO THEY PULL?

Load lifted on dynamometer	Equals starting wagon load on pavement	Equals 14 inch plows working 6 inches deep in stubble ground
1,800 pounds	23,400 pounds	Four plows
2,400 pounds	31,200 pounds	Five plows
2,900 pounds	37,700 pounds	Six plows
3,300 pounds	42,800 pounds	Seven plows
3,600 pounds	46,700 pounds	Eight plows
3,900 pounds	50,600 pounds	Nine plows

Lyle Orwig

STEVE LANCE works 40 pulls a year with his truck-mounted dynamometer. Taking the place of a weighted stoneboat, Lance says this machine gives a more uniform pull. Lance has seen a team pull the equivalent of a 50,600 pound loaded wagon with this machine.

Frank Lessiter

J.C. Allen and Son

Frank Lessiter

REFINEMENTS have been made in dynamometers over the years, although the basic principle has remained the same. Pull officials in this 1926 contest dressed differently than do their modern-day counterparts.

HIS WHISTLE to his lips, Lance watches intently. Once the team has pulled the required 27½ feet, you'll hear the loud shrill of the whistle signal the team to stop.

147

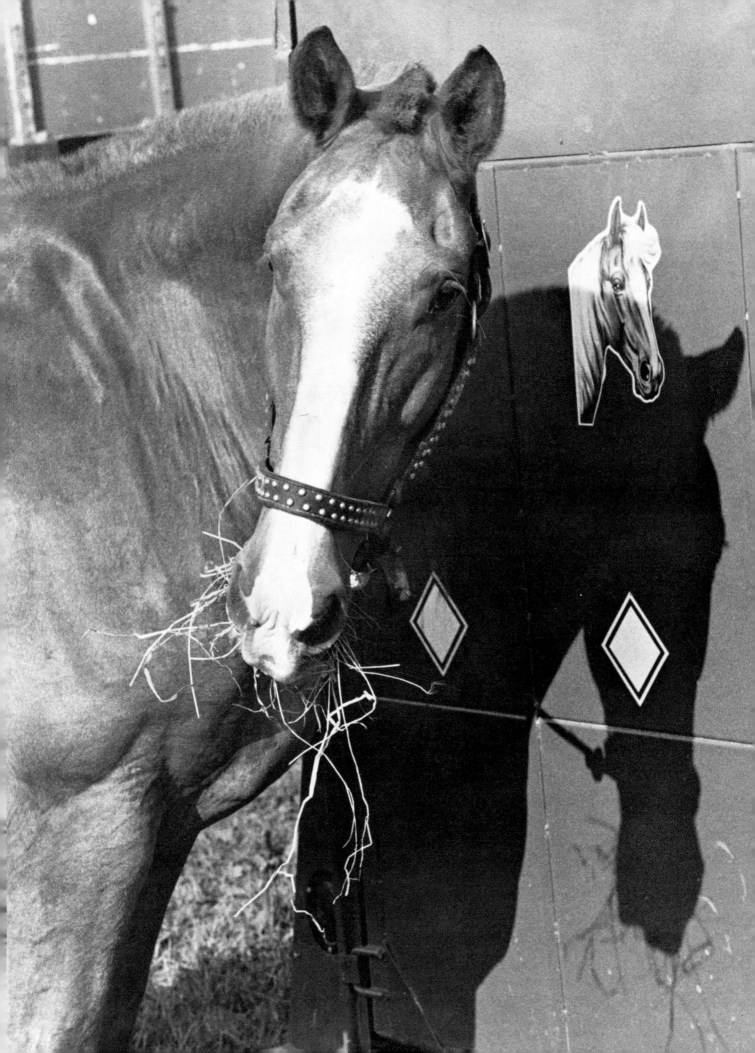

A Champ Shares His Pulling Secrets

WHETHER IT'S a horse pull on sod or dirt, you may hear the announcer include the name "Grass" when he announces the contest winners.

That will be "Grass" as in Marshall Grass. He's the Blair, Wisconsin, dairy farmer who has won more first places than he can remember in the more than 40 years he has been pulling horses.

While Grass has had a tremendous horse pulling career, he'll quickly tell you his greatest draft horse thrill occurred on September 18, 1974, at Centerville, Michigan.

That's the day his Dick and Prince team of Belgian geldings—topping the scale at a combined weight of 4,700 pounds—helped Grass reach the highpoint of his horse pulling career.

But it was a struggle. Both the Grass heavyweight team and another team owned by Harry Roehl of Smith Creek, Wisconsin, pulled 4,400 pounds in the contest. But the difference was that the Roehl team only pulled the load 9 feet, 5 inches while the Grass team pulled the load 15½ feet.

That difference of 6 feet, 1 inch—close to the height of Marshall Grass—was enough to make the Grass team the National Heavyweight Champion horse pulling team for 1974.

If the team had pulled the 4,400 pound load 27½ feet on that early autumn day at Centerville, it would have been a new world's record. But to qualify for the world's record, a load has to be pulled the full 27½ feet. The present world's record of 4,375 pounds was set in 1971 by Fowler Brothers who farmed down

"THE FIRST THING I always look for in a pulling horse is the animal's head," says veteran puller Marshall Grass. "I like to see a roman-nosed pulling horse. I don't know why I tend to look at the head of any new horse first, but I always seem to."

Frank Lessiter

Frank Lessiter

149

near Montgomery, Michigan.

It's easy to see why that was a memorable day in the Grass horse pulling career. Yet Marshall is quick to admit to many great days spent pulling horses or training horses to pull.

At the 1976 World Championship Horse Pull, Grass beat out 18 other teams to capture first place and $1,000 in the heavy-

> "Competitive spirit is an essential for good pulling horses..."

weight class. His pull: 3,900 pounds on the dynamometer!

This is equal to pulling 50,600 pounds of weight with the team—that's nearly 11 times the combined weight of the two horses! Or put another way, this is the same as pulling nine 14-inch wide moldboard plows 6 inches deep in stubble ground. Just picture that! That's *real horsepower*.

Marshall Grass talks about his Dick and Prince team with the pride many a man reserves only for his children. In fact, Grass readily admitted he treats those horses like kids... and that's no offense to his son, Allen, who helps crop 350 acres and milk 60 dairy cows on the family farm.

Of course, Allen also has the horse pulling bug, with a middle-weight team of his own that has done pretty well in a number of pulls.

4,700 Pounds of Power

The horse named Dick—the biggest part of the team at 2,400 pounds—was purchased at three years of age from a Minnesota farmer.

"This horse was really something," says Grass. "If you put him on any kind of load, he had the determination to pull it. He loves to pull and finds it a challenge—I think he would try any load regardless of the weight."

The other horse, Prince, came from a Michigan farm, but was bred in Minnesota. The scale needle flies up to 2,300 pounds when he lumbers on.

Prince had already appeared in about a half dozen pulling contests when Grass bought him. But Dick had never been pulled at all.

"One thing about the temperment of that team was that you wanted to try to keep them as quiet as you could before a pull," explains Grass. "They got excited just like kids. They could smell a horse pull coming and were always ready. They had the competitive spirit—and that's an essential for a really good pulling horse."

Lots of Good Pullers

Grass has had a number of good pulling horses over the years. Besides Dick and Prince, he quickly recalls another horse called Prince... an old gray horse called Jess... and another named Dick he had back in the early 1950s.

"That Dick was a big bay horse we pulled more than 300 times in seven years," recalls Grass. "He was only beaten 11 times.

"Then one day I walked out into the pasture and found him lying dead. We knew he had heart trouble, but it was still a real shock to lose him."

Years ago, Grass started out like many green recruits to the

> ### BIG FEED BILLS
>
> THE Grass team of Belgian geldings split a 40-pound bale of hay a day. And every time Grass walks into the barn, he tosses them a few handfuls of a home-mixture of corn, oats, molasses, vitamins and minerals.
>
> At home, they are turned out for exercise each night in the summer after the sun goes down and the flies disappear. Grass also hauls green-chop that is fed to the horses in the exercise lot at night.

DICK AND PRINCE fill up the fifth wheel trailer Marshall Grass tows behind his pickup.

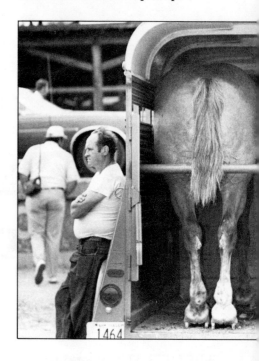

Frank Lessiter

FRIENDLY JOSHING is a normal ritual among the horse pulling fraternity.

Frank Lessiter

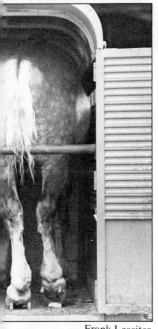

Frank Lessiter

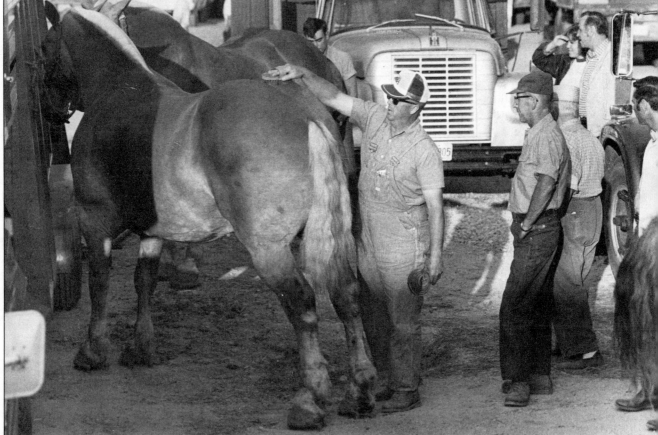

Frank Lessiter

LAST MINUTE harness adjustments are made by Marshall Grass before his team goes into competition.

WELL-CURRIED HORSES are a sense of pride to most teamsters. While others view the immense size of these two giants, Marshall Grass works to have them looking their best for the upcoming pull.

151

DRIVING SKILL is something gained with experience. "I've lost a few horse pulls myself simply by over-driving the team," says Marshall Grass.

A WINNING PORTRAIT of the team with which Marshall Grass won the National Heavyweight Horse Pulling Championship. They pulled 4,400 pounds.

Frank Lessiter

horse pulling contest. Busy with farming in the early days, he only entered three or four pulls a year. But then as he developed a few good pulling horses, the competitive horse pulling fever took hold.

"I remember we took in 33 pulling contests one year," he recalls "In 1944, we were gone from the farm a week and pulled in six contests.

"You could hit a different horse pull every day of the week in those days. Now the horse pulls pile up on the weekends"

The number of horse pulls Grass enters each year has now dwindled to around a dozen. But he hits most of the big contests without fail every year from spring through fall unless one of his horses is injured or needs some rest.

Give 'em a Chance to Pull

Grass likes to buy young horses that appear to have the stamina and power for pulling, then train them himself.

If you were to start out with 100 horses, Grass figures you would wind up with only one or two really top-notch pulling horses. Many other horses would turn into average or good pulling horses, but only one or two would become really *top* pulling horses.

"I would rather have two green horses to train than a green horse and a horse that is fully-trained," says Grass. "When we have only one green horse, we hook him with an old experienced farm horse that has never pulled. He makes the best teacher of any horse I've ever

---

"GIVE ME QUACKGRASS"

GROUND conditions vary greatly from pulling contest to pulling contest. And the ground can have a big influence on the kind of load a team can pull.

Marshall Grass prefers quackgrass sod with just a little moisture in it. Yet most horse pulls are held on dirt.

---

been able to find.

"He gives a young horse a real chance to pull on a wagon or sled. And regardless of what the new horse does, the old horse is always in there pulling."

Grass trains a new horse on the hay wagon, sleigh, manure spreader or cultivator. Grass uses his horses to feed corn silage to dairy cattle from a sleigh during the winter. And he also works new horses on a sled on sod during the summer months.

Grass tries to start new horses on a half-filled load of silage on sod. "We always try to have a new horse start on a load he can pull—that builds his pulling courage," says Grass.

"We have the horses do just about everything and anything around the farm. If they can't be used to work on the farm, they certainly can't pull.

"Winter is usually when we work new horses. And we do it every day. If a horse shows the potential of making a pulling horse, we keep after him.

"But I can remember horses

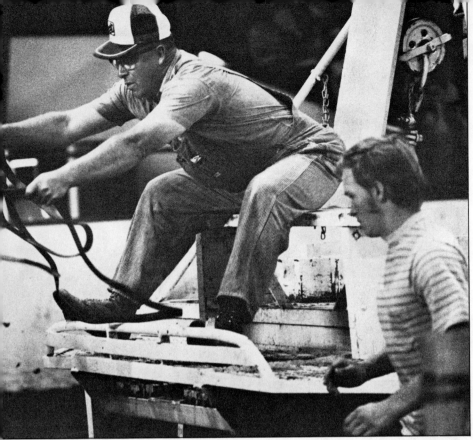

Frank Lessiter

we pulled every day all winter. When spring came, we still didn't have any good pulling horses."

Hard farm work helps a horse develop into a real puller. According to Grass, "You can tell when you are feeding cattle with horses whether they want to pull or not. A successful pulling horse needs a burning desire to pull.

"You can tell by the way they handle a sled loaded with silage on the crushed rock shoulder

---

"You could hit a different horse pull every day of the week..."

---

along the highway. If they walk easily, you know they can handle it.

Wants Different Horses

"You just have to practice until a young horse gets the hang of pulling. He has to learn to get under a load. He may be fast or slow—it doesn't matter as long as his pulling partner works at the same pace."

Selection of a pulling horse varies according to the personal preferences of every puller. "But the first thing I always look at on a horse is the head," says Grass "I like a roman-nosed horse. I don't know why I first look at the head of a horse, but I do."

Grass also wants a long-bodied horse showing good muscling over the kidney and stifle. Feet and legs have to be good if a horse is to stand the tremendous pressure of pulling. There's an old saying in the pulling trade: "No feet, no horse."

He also wants a horse with plenty of get up and go, good temperament and desire.

The bigger the horse, the better. "I like to select a horse that is three or four years old, so I can see how big he is really going to get," suggests Grass. "A lot of horses are deceiving at an early age. They look like they are going to get big, but then they don't."

But regardless of how you select a pulling horse, Grass admits, you aren't sure of what you

really have until you pull him for a few months.

Man Behind the Horses

After you've picked out the proper horse, trained him perfectly and had a farrier shoe him properly, you've still got to train yourself or some other teamster to handle the team.

Grass believes the main thing a driver must watch for is not to

---

"It was enough to make them the National Heavyweight Champion horse pulling team..."

---

over-drive the team during a pull. It's a nervous time for both team and driver.

Grass says he speaks from experience when he tells newcomers to watch the way they handle a team. He's lost a few pulls himself simply by over-driving a team.

Grass adds that every horse must be handled differently, and nearly every driver operates in his very own way. "A driver may call on the team or on one horse to pull harder," he explains.

"Some drivers whistle, chirp or even cuss. Other drivers will tell you that if they said a word, the team would stop dead in its tracks during a pull. Some drivers keep the lines hanging loose; others hold them tightly."

Crowd noise can also be a factor, but you can't really do much about it. A year or two ago, the Grass team stopped dead in their tracks during a pull as a Fargo, North Dakota, crowd clapped and cheered loudly. Luckily, they had just passed the 27½ foot mark that made the pull official.

Grass is mighty proud of his record over the years in both pulling and training horses. At one time, he knew of 27 horses entered in various pulls that he had owned at one time or another.

That's a record any horse puller can look to with pride.

153

United States Department of Agriculture

Jim Pottorf, MFC Services

Horse Laughs

DIDJA HEAR the one about the draft horse that got hitched to a donkey?? Yeah, he made a real jackass of himself.

That may not be the kind of horsey humor that turned on these draft animals, but they appear to be laughing ear to ear just the same—especially that mirthful mare below!

Actually, no one knows what tickled their hides and brought out these expressions for the photographers. Horses can have secrets, too.

Edward Druck

155

Meet a Blue Ribbon Horse Man

HAROLD CLARK could paper his whole house with the blue ribbons he has won during 54 years of showing draft horses.

Actually, that's probably an under statement. He could likely cover the walls of his *whole house* and at least half the horse barn with the *thousands* of blue ribbons he has won over the years.

Some people believe that Clark, who now lives at Millersburg, Indiana, has probably won more blue ribbons than any other man in the history of the show ring. And Harold will modestly tell you there were years when he won so many blue ribbons that he was sort of embarrassed by it all.

One of those years was 1971. Out of the five major Belgian shows he exhibited at, Clark came away with *64 blue ribbons and 23 championship awards!* His horses were named grand champion stallion and mare at all five shows.

Clark grew up in the draft horse business on the family farm located near Sheldon, Illinois. His dad farmed with horses and had always wheeled and dealed horses as a sideline.

"Horses always used to be shipped into our area from Wyoming and Montana in February for sale at a local auction," says Clark. "My dad would later buy

WHETHER IT was a circus parade, a horse show or a judging chore, Harold Clark was always ready for action. Clark knew all breeds of draft horses—having shown horses for 54 years. At five big shows in 1971, his Belgians took 64 blue ribbons and 23 championship awards.

Frank Gardner

Jos. Schlitz Brewing Co.

HAROLD CLARK WORKED with Belgian, Clydesdale, Percheron and Shires. Spending eight years with Percheron horses here at Indiana's Conner Prairie Farm, Clark showed the three best mares in Chicago for seven straight years.

J.C. Allen and Son

the horses that nobody else wanted at the auction. That way we would usually get 25 to 30 head to break and train each year for the sale the following February."

Clark recalls many fond moments working with those horses on the home farm. "We used to plow when I was a kid with 12 to 16 horse hitches," said Clark. "You would drive a few of the horses—the rest of the horses were just tied into the hitch. They helped with the pulling, but they just followed the lead of those that were being driven.

"We could pull four 16-inch plow bottoms with a 16-horse hitch and three 16-inch plow bottoms with a 12-horse hitch. Our soil in Illinois was pretty heavy, so you always needed lots of horsepower."

He also drove his biggest hitch during those teen-age years . . . 36 horses hitched to a road construction excavator. He still fondly recalls that experience.

Went to Greener Pastures

Clark left the home farm on February 8, 1920, to work for the Holbert Horse Importing Company at Greeley, Iowa. "I used to read that Holbert ad in the farm magazines and it looked interesting," he recalls. "So I went out there and joined up."

The morning of his very first day on the job the boss told him to deliver a horse to Masonville, Iowa. "I've always meant to look it up on a map but I still haven't," says Clark. "It was 40 miles. I just walked that horse. It was around midnight when I got there."

Holbert's, he explains, imported Belgian, Percheron and Clydesdale stallions from their native European lands and sold them throughout the United States and Canada. They always had 100 to 200 stallions for sale at any one time.

Each man working for the firm cared for 25 horses. He had to have them looking their best each day at 9 a.m. when prospective buyers arrived to look over the selection.

Buyers would keep coming

J.C. Allen and Son

BIG OR SMALL, the Clarks enjoyed good horses. Clark's three-year-old nephew, Jimmy, posed with Conner Prairie's great Percheron stallion Ostralien.

and keep looking over the horses until dark. Then the barn crew could feed and bed down the horses for the night . . . and get some rest themselves before early morning quickly rolled around and they had to start the whole procedure over again.

Holbert's also had branch stallion sales barns in other parts of the United States and Canada. Being a bachelor at this point in his life, Clark spent lots of time transporting horses to these other distant barns and working at different branch sale barns.

"I also remember when I was 19 and they sent me all alone to New York City to take 120 draft horse stallions off the boat," says Clark. "I brought them back to Iowa by rail. I had my hands full for a young lad."

It was with Holbert in the late 1920s that Clark showed his first draft horse champion at Chicago's International Livestock Exposition.

One Stallion, Three Cows

"One thing I remember well while working for Holbert's was that if we sold a horse to another farmer located 50 miles away, we always shipped the horse by rail," says Clark. "But if the farm was less than 50 miles away, you *led* the horse to his new owner!

"You got up early in the morning when you had to lead a horse 40 to 50 miles, let me tell you. And sometimes I think they underestimated those distances a few miles. I'm sure I walked a few horses 55 miles to their new owners."

In 1928 Clark joined Frank Huddleston at Webster City, Iowa, a breeder of Shire draft horses. This was his first association with water troughs and manure spreaders, the latest labor-saving devices.

Moving next to the farm of Harry Stamp, Roachdale, Indiana, Clark had charge of 100 to 150 Belgian and Percheron horses. It was here he began to

159

do all of his own horse shoeing.

"They just couldn't shoe a horse to suit me," he recalls of the area's blacksmiths, "so I did it all myself." And he learned to do much of his own veterinary work because his boss Stamp "liked to keep the vet bill down."

Stamp was on the road most of the time selling horses. This was during the depression years, yet Clark remembered a good draft horse stallion still brought $1,000 to $1,200.

"Horses were worth more money than any other farm product in those days," says Clark, who stayed at the Stamp farm for two years. "A good thumb rule was that a stallion was worth as much as three good cows. And I think that's pretty much still true today."

Best Mares for Seven Years

His next eight years—his most magnificent in the show ring—were spent at Conner Prairie Farm, Noblesville, Indiana, where he showed the best three Percheron mares at Chicago's International Livestock Exposition for seven straight years.

"There were some tough battles in the show ring in those days," Clark notes. "They were always out to beat a winner."

In 1937, Clark exhibited "Lancinante", a Percheron mare, to many championships across the country, including the International in Chicago.

"It was tough to bring her back in 1938 after she'd won so much," he remembers. "That 1938 win at Chicago brought me the closest I ever came to crying. After I squeaked out the win, somebody stuck his hand over the rail to shake hands with me and I expect my face was a little wet."

The following year Clark learned an invaluable lesson which he vividly recalled many times over the later years. Conner Prairie Farm bought another Percheron mare, Julie, at Clark's near insistence. The farm's owner, Eli Lilly, didn't

SEVERAL years were spent by Clark at the Harry Stamp farm near Roachdale, Indiana. Stamp's Belgian stallion, named Victor, was grand champion at a number of big draft horse shows in 1931.

J.C. Allen and Son

THIS MARE once brought Harold Clark close to tears in the show ring at Chicago. Lancinante won many championships in 1937, including the big one at the International Livestock Exposition. Clark came close to crying when she repeated in 1938.

Cook and Gormley, Eli Lilly and Co. Archives

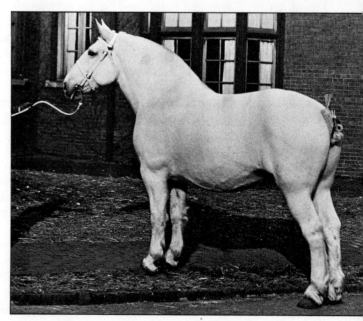

JULIE was a mare that Harold Clark told farm owner Eli Lilly was good enough to win grand champion Percheron mare honors at Chicago. And Clark was relieved when she did it in 1939.

Cook and Gormley, Eli Lilly and Co. Archives

Cook and Gormley, Eli Lilly and Co. Archives

FOR SEVEN straight years, Harold Clark showed the best three Percheron mares at the International Livestock Exposition in Chicago.

feel she was of Lancinante's caliber.

"I told him she was good enough to be champion at Chicago," Clark says. "But there can be quite a difference between being good enough and being champion.

"When anybody came to the farm that year Mr. Lilly always had me bring the new mare out and he would say, 'Mr. Clark says she'll be champion at Chicago.'"

That increased the pressure, and Clark admits to a little nail biting and extra care for Julie that summer before the show circuit got started. "And I learned from that experience to keep my mouth shut," he says. "I expect when she was named champion at Chicago I was as relieved and tickled as anybody."

On June 5, 1942, Clark moved to Meadow Brook Farms at Rochester, Michigan. It was owned by Mrs. Wilson, the former Mrs. John Dodge of auto fame. Clark had charge of some 70 Belgian breeding and show horses. Additional horses for field work were owned and cared for by Meadow Brook's farm department.

Mrs. Dodge gave the farm to Michigan State University and moved the horse breeding operation to a Fowlerville, Michigan, farm in 1960. After her death, Premier Corporation bought the farm and the horses in 1969.

Clark continued to run the Belgian horse operation until he and his wife, Ruth, bought all of the horses and moved to Millersburg, Indiana, in 1973. Due to poor health, he later had to sell most of the horses. But Clark continued to hold an interest in a few choice horses around the country.

We Love a Parade

While Clark always enjoyed a "thrill a minute" in the show ring, you wonder after visiting with him at length whether driving wasn't his greatest delight. It's no easy job to put together a large hitch, and Clark was just as successful at doing that as in showing individual horses in the ring.

It takes real knowledge to

161

select the right animals for the wheel team, the swing team and the lead team (or eight horses instead of six for big parades, such as the Milwaukee Fourth of July Circus Parade that was held annually from 1963 to 1973).

In addition, you have to figure out the proper way to bit each horse. You must also know how to design, cut and fit the raw steel shoe that will help them maintain the perfect gait.

Surprisingly, Clark never showed any draft horse hitches until he arrived at Meadow

---

"One year we showed for 17 weeks. Only once all year did we load or unload when it wasn't raining..."

Brook Farm. Part of the reason was that, up until about 1945, most show hitches consisted of Clydesdales, and he had been working primarily with Belgians, Percherons and Shires.

Clark has fond memories of exhibiting a six-horse hitch at Toronto's Royal Winter Fair in 1946, the first year the show was resumed after being closed for seven years due to World War II.

"We had the first United States six-horse hitch to ever win in Canada in 1946 and we won again in 1947," says Clark. "The Canadian horse breeders said we would never win again at Toronto after we won in 1946. But we were lucky enough to repeat our win the following year."

Greatest Thrill Ever

Clark has had numerous thrills in the horse business over the years. "But one of the biggest has to have been in 1946 when we won the six-horse class in Toronto and then won in Chicago the next month," he recalls.

"The thing that made this especially unique was this was the *only* six-horse hitch in those days that won at the big shows which had all been bred and raised on the same farm. The other hitches had been assembled

162

by buying horses from various farms."

During the Meadow Brook days, Clark and the big Belgians used to take in around nine big shows a year. They would start with the Illinois, Indiana, Ohio, Wisconsin and Michigan State Fairs. Next came the National Belgian Show, the Toronto Royal Winter Fair and then the big grand finale was the International Livestock Exposition in Chicago.

They used to take 12 to 20 horses to each show. While they moved by rail in earlier days, trucks were used later on.

One of the bright spots of older-day railroad transportation was being able to ship purebred livestock or livestock used for exhibition at half the normal freight rate. "From Greeley, Iowa, to Chicago, you could get an express railroad car that would hold 20 horses for a roundtrip cost of just $80," said Clark. "Today, at a trucking cost of 70¢ a mile that 400-mile roundtrip would cost you $280.

"When shipping horses by rail, the railroad always gave you three days at the fairgrounds to unload without charging extra for the car. Then you could have a car delivered to the yards three days earlier than you needed it. So with a little luck, I was able to keep the same railroad car

---

BUDWEISER BELGIANS?

HAROLD CLARK always chuckles when he recalls going to shows in any big city. The first job would be to move in the horses and the harness.

About that time city folks wandering down the aisles would say, "Hey, the Budweiser horses are here!" These folks figured every team of horses they saw belonged to Budweiser. "That showed me what big horses and advertising can do," says Clark.

Premier Corporation

throughout an eight- or nine-week show season without having to pay extra for the car sitting idle at the fairgrounds."

Clark would carry a dozen horses in a box car. They would not double-deck the car since the big horses needed plenty of head room. Hay, straw, harness, feed, tack and cots would be placed in the center aisle opposite the big doors.

Clark made a number of long trips by rail riding right back there with the horses. In 1926, he made a five-day trip from Peoria to a 150-year United States anniversary celebration in Philadelphia. That same year, he made a seven-day, 1,750-mile trip from Chicago to Portland, Oregon. That's a lot of time in a box car with a dozen horses!

LOVING PARADES, Harold Clark drove six or eight horses. His greatest thrill came when his six-horse hitch swept top honors at Toronto and Chicago way back in 1946.

"The year 1926 was not the best year weather-wise for shipping horses," recalls Clark. "We showed horses for 17 weeks in 1926. I remember we only loaded and only unloaded once all year when it wasn't raining at various fairs."

"We Were Darned Scared"

One of the biggest scares in Clark's life happened in a box

"Seven cars in front of our box car left the rails and turned over..."

---

car. He explains it best: "My brother and I had a box car load of horses coming back to Indianapolis from a horse show out East. We were standing in the doorway looking out toward the front of the train when it happened.

"Right before our eyes, we saw seven cars in front of our box car leave the rails and turn over. There was a coal car just in front of us. I said to my brother that if the coal car leaves the rails, we've got to jump. We didn't want to be tumbling around upside down in a car full of big horses.

"That coal car was the first car of the rest of the train that didn't leave the tracks. We came that close!

"There was a hobo in the coal car and he had been watching, too. After it was all over, he walked down the tracks shaking his head saying he would *never* hop another freight as long as he lived."

What's it all add up to... Clark's long career in the draft horse business? Probably a whole book full of stories all its own. But the thing Clark values most from all those experiences is the *hundreds* of good friends it's given him.

"Nobody has ever done anything as long as I have and had so much fun and made so many friends," concludes Clark. "Good friends rather than a lot of money is what made the draft horse business a *great life* for me over all these years."

163

ONLY STOLE ONE HORSE

"FIRESTONE" was a gelding anyone would have been proud to own regardless of their special breed of draft horse. And if you know anything about Belgian horses, you've probably already heard about this horse that was one of Harold Clark's all-time favorites.

At Meadow Brook, draft horses for working the farm were cared for by the farm department rather than the Belgian breeding operation Clark headed up.

"One day I saw this gelding, Firestone, in the farm department's pasture," says Clark. "I was immediately impressed by him, and I went out and caught him and brought him to the show barn. He was the first and last horse I ever 'stole.' "

Born in 1938, the horse turned in a fantastic show ring record over the years. He was champion gelding three times at Chicago's International Livestock Exposition. He won champion Belgian gelding honors three times at the Toronto Royal Winter Fair, and was grand champion over all breeds once in Toronto.

When Firestone finally retired from the show ring, he had won *58 championships!*

He was sired by Progress, the noted stallion which sired a great many of the winning Meadow Brook Belgians. Weighing 2,200 to 2,300 pounds, Firestone was a strong, well balanced horse with plenty of character.

With a massive, thick, deep body, Firestone looked every inch a draft horse. He was without a doubt one of the greatest draft horses of all time.

Abernathy Photo Co.

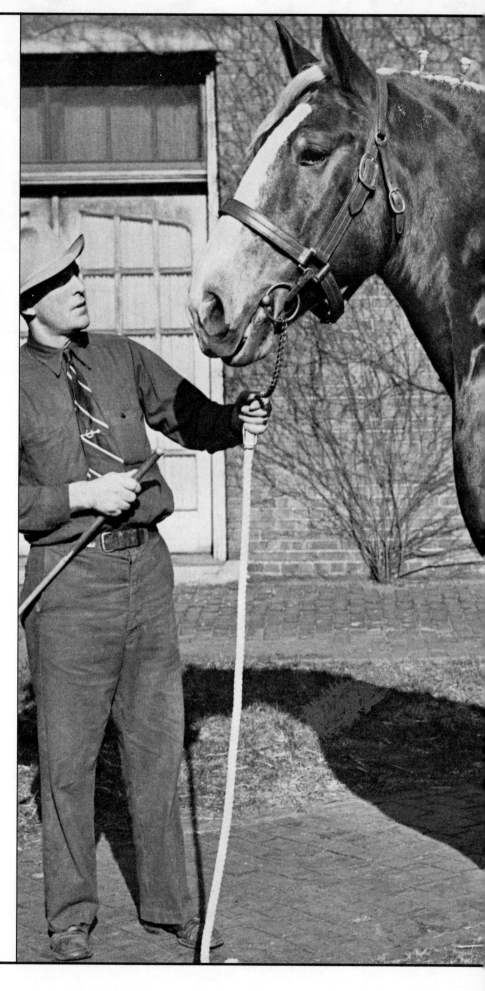

Premier Corporation

MANY DIGNITARIES had a chance to ride on the wagon with Harold Clark over the years. Included were these three former astronauts and their wives in a Michigan State Fair daily parade.

THE LION AND MIRROR bandwagon was pulled by Harold Clark's eight-horse hitch for a number of years in the annual July 4th Milwaukee Circus Parade.

Harold Cline

Robert Nandell, Milwaukee Journal

It's a Different Parade

Saturday afternoon, when Harold Clark came rumbling down Milwaukee streets with his eight-ton circus wagon loaded with red bandsmen tooting "Gentry's Triumphal" or "Quality Plus," this writer sat at Harold's side, and this is what he experienced from that seat just to Harold's right:

OLD MEN grinned until you'd think their teeth would fall out. Fat ladies in lawn chairs giggled and jiggled. Everybody's daddy brought their cameras up, clicked and whirred... and there was at least one cameraman who is going to have an extreme closeup of his index finger because, there it was, right over the lens!

When Harold's eight big Belgian draft horses came strutting, clopping and jingling down the street, and bandleader Fred Parfrey gave the signal to strike up the music, you could *feel* the cheer and heart-swelling coming up out of the crowd like steam from the calliope.

And the applause came, splattering up like a reverse rain shower, and making the guys in the band feel like playing even more and louder.

This Band Had Class

This wasn't just any old pickup band, either. This band had *class.*

Right in front was Merle Evans, who led the Ringling Bros. and Barnum & Bailey circus bands for more than 50 years. He was blowing the silver cornet with his ageless lips. Blowing it right back at him on the next bench was Jimmie Ille, his successor as leader of the Ringling band.

The response to that crazy arrangement of horseflesh, harness, wagon, men and music by the parade watchers was the sort of thing that grabbed you by the throat and made warm tears come to your eyes. The hair still rises on my arms now as I vividly recall it.

Kids! Oh, there was a population problem there in the beer town that Saturday. But from the wagon seat up there beside Harold, it didn't look like any problem.

It was a bright-eyed, sticky-fingered, freckle-faced delightful flood of young smiling faces... with a few teeth missing, dirty knees

From the Driver's Seat

and every last young soul soaking up the size of those clomping Belgians. They were all bobbing with the music and waving and laughing.

Corners and Miracles

Harold was pulling at those eight reins—I wish you could have seen how his feet were braced against the footboard! Every time he went around a corner it had to be some kind of a miracle.

But Harold didn't seem to think so. All in a day's work for a teamster who once drove 36 horses on an excavator. But his face couldn't hide the thrill just the same.

Eight horses? That's nothing when you have a team like Harold was driving. There was Steve and Rowdy making up the lead team, Rock and Jim on the second swing, Mark and Rebel on the first swing, and Bruce and Tony right next to the wagon as the wheel team.

Each horse weighed a ton, and Harold only goes maybe 170 pounds. But the eight leather straps are wound tightly between his fingers—one line going to each horse. His arm muscles are taut, his legs are braced, and there's no doubt that he is in full charge.

"Every time you hook them up, one of them will act a little different than the last time you had them together," Harold had told me earlier.

This day, he was having a little problem with Jim. For some reason Jim is riled up and has decided to work himself into a sweat. And he is making one of Harold's fingers numb with his ambition.

"I guess he just decided to pull today," Harold says, "and that makes it tough to drive them."

Horses Couldn't Hear

A couple of times, Harold had to ask the band to stop playing so the horses could hear his commands. He gets the big band wagon moving by whistling and stomping his foot. His whistling sounds just like the piccolo playing, but the horses mysteriously can tell the difference.

At least that's the way it looked from up there on the wagon seat next to Harold and Merle and the boys. —*By Bill Stokes. Reprinted from the Milwaukee Journal.*

FIELD DAYS always brought out crowds to study the best stock a draft horse breeder had to offer. The crowd would listen intently as a college prof went over the virtues of the farmer's breeding battery.

COLT CLUBS were popular in the days when the draft horse reigned superior. Back in 1928, 26 teamsters and their colts paraded down this small town's main street on the way to their annual colt club show.

J. C. Allen and Son

J.C. Allen and Son

Show Ring Fever Gets in Your Blood

Frank Lessiter

SHOWING DRAFT HORSES is something that gets in your blood and you can't ever get it out, says Justin McCarthy of Parnell, Michigan.

This veteran draft horse breeder has been raising and showing horses for more than 60 years. While retired from active farming now, he still takes in a few horse shows each year and always lends a hand whenever and wherever he can.

McCarthy recalls many exciting instances in the draft horse business. He showed from 1913 to 1950, usually making seven to ten fairs each year.

International Was Big Show

"But the granddaddy of draft horse shows in the old days was always the International in Chicago," recalls McCarthy. "It was the World Series of the draft horse shows in those days.

"And you only took your best

RAISING AND SHOWING horses for over 60 years, Justin McCarthy says the thrill of working with big horses is something he always enjoyed.

169

YOUTHFUL JUDGES used to place draft horses as well as cattle, hogs and sheep in their contests.

J.C. Allen and Son

horses. If you usually kept ten horses in your show string, you would probably only take the three best ones to Chicago."

While they travelled mostly by rail from fair to fair in the old days, McCarthy always walked his horses the 13 miles from home to the Grand Rapids fairgrounds. "There would be two of us and we would each tie five horses together," he remembers. "We would usually take one saddle horse along and take turns riding him. We could go those 13 miles with the horses in about four hours."

But by 1920, they were trucking the horses from fair to fair.

Horses used to be exercised at the fairs out on the race tracks. McCarthy would lead several horses with the saddle horse each morning around the race track for two miles. That gave the horses about the right amount of daily exercise.

McCarthy's family had always been interested in horses of one kind or another. His grandfather raised jumping horses in Ireland and also did surveying. "But he got sick of paying taxes all the time over there and came to the states," says McCarthy.

"He used to tell how one man would come along, collect the yearly taxes and leave no receipt. Then another man would come around a month later and collect the yearly taxes again. He got tired of paying his yearly taxes three or four times annually, so he said the heck with it and came to the states in 1841."

After arriving, McCarthy's grandfather got a job with a government crew surveying for

---

"Okay, as long as my dad never sees me leading a Shorthorn instead of a Belgian . . ."

---

the railroad at Maumee, Ohio. But when the survey job was completed, there was no money to pay him. So he was offered a "special deal" on some land in the Grand Rapids area. "He came up here, took a look, bought 160 acres for $1.50 per acre and started farming," says McCarthy. "We later expanded the size of the farm to 385 acres."

McCarthy used to "trade" horses a great deal. One year he would have 60 draft horses. The next year he might be down to 20 head. And the next year he might be back up to 50 horses.

McCarthy started out raising Belgians and Percherons. "The Belgian mares did the best job for me raising colts," he recalls. "So I got rid of the Percherons after ten years.

"I had better luck with Belgians. They seemed more rugged and were raising a colt for me every year. I always had to babysit my Percheron colts. But other breeders did okay with them.

"I had a Clydesdale mare once. But I quickly decided that keeping the long feathered silky hair on her legs clean and in good condition was too much laundry work for me. So I sold her."

McCarthy finally quit showing and sadly sold most of his Belgian horses in 1950. "That was a time when you couldn't even give a good horse away," he recalls. "We had made good money from draft horses over the years, but we couldn't at that time."

Frank Lessiter

Frank Lessiter

"I LEARNED to braid tails when I was 13," recalls Justin McCarthy. "It's only a wild guess, but I bet I've braided 3,500 tails and rolled 500 manes. An old pro showman who drank a lot taught me once at a fair how to roll a mane. I've never forgotten."

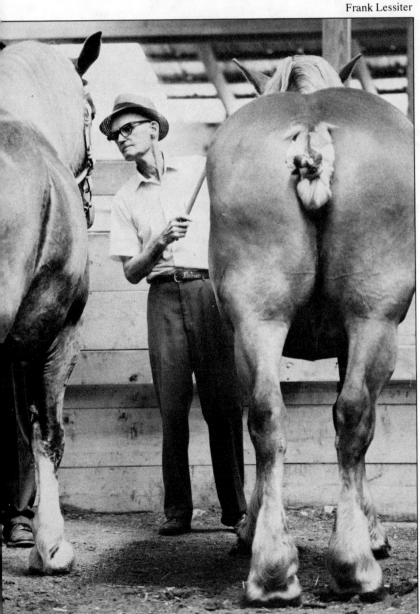

Frank Lessiter

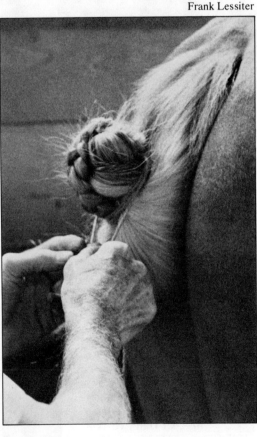

Frank Lessiter

ONLY THE CREAM of the draft horse crop was shown at Chicago each year. Justin McCarthy remembers seeing 65 two-year-old stallions in one class. The biggest class he ever showed in at Chicago included 31 yearling mares. If a horse wasn't up to snuff, the two judges would turn a horse back to the barn before it ever entered the show ring.

171

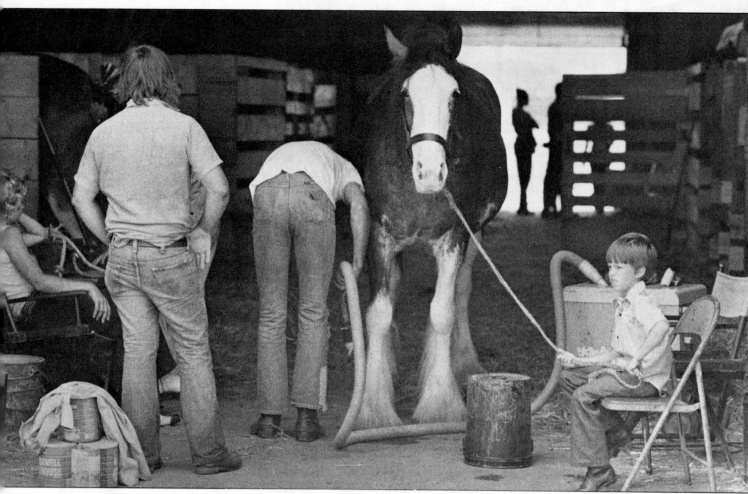

Frank Lessiter

MORE THAN CARPETS get a thorough vacuuming by draft horse owners. Many give their horses a last minute cleanup with the vacuum cleaner before heading for the show ring. Sitting a restless wandering young boy down on a folding chair keeps him in sight... and lets him do double duty by keeping a tight rein on this Clydesdale colt.

Frank Lessiter

"**WALK HER**," the show judge has just told Belgian breeder Dan Creyts of Lansing, Michigan. After studying a standing horse closely, most judges want to see how the horse looks in motion. Quite often, a showman will be asked to have his horse perform at both a walk and a trot for the critical eyes of the judges.

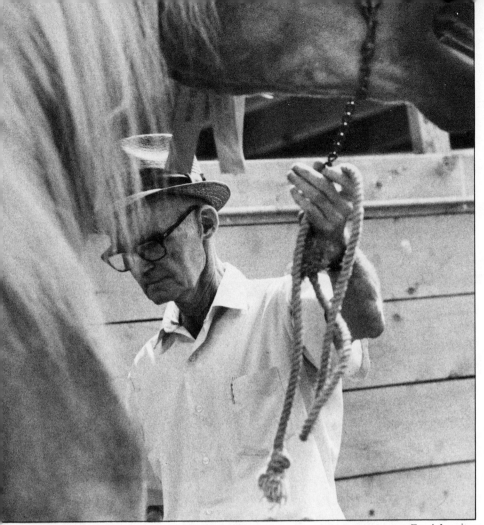

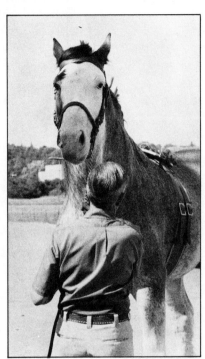

THE GIANT HORSES get so tall that they often tower over the heads of the showman—like Belgian man Justin McCarthy of Parnell, Michigan, at left . . . or Clydesdale breeder Myron Avery below from Marshall, Michigan.

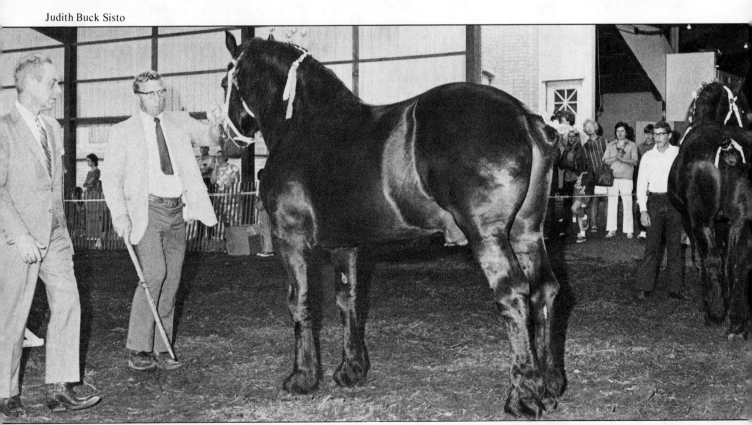

FIRST OR SECOND PLACE seems to be on the mind of the judge as he carefully looks over this Percheron horse.

173

MANE ROLLING took time and wasn't something where real skill was learned overnight. The veteran horseman rolling this mane is Elmer Taft of Indiana's Lynnwood Farms.

ON AN early fall day in 1928, this yearling stud class was judged at Indiana's Wabash County Colt Show.

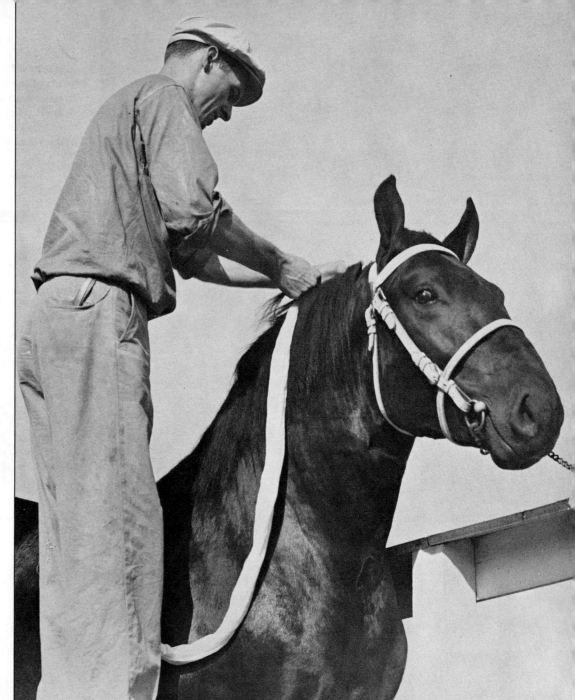

J.C. Allen and Son

J.C. Allen and Son

J.C. Allen and Son

J.C. Allen and Son

SOME 25 Belgians made up this impressive class at an Indiana State Fair in the days when competition was fierce.

HORSEFLESH judging was learned by many 4-H Club members in earlier days. It was more important then to be able to cite the differences between a good pair of mares than the differences between a Model T and a Model A.

OBSTACLE course driving shows the talents of both a teamster and his team. It takes patience and practice to ease a team through a space only inches wider than the wagon.

Judith Buck Sisto

OUT FOR some racetrack exercise, this team is "worked in line" instead of the more normal way of hitching two horses abreast.

Judith Buck Sisto

Judith Buck Sisto

Frank Lessiter

IN NEW England, teams often compete in "worked in line" classes as well as the more popular two abreast team class.

ARMY VETERAN William Young of Clair, Michigan, has worn his watch on his shirt collar buttonhole since his days in the service. "I saw many an Army man lay his watch down on the sink while washing his hands," he explains. "When he finished, the watch was often gone. I've worn my watch through my shirt buttonhole ever since."

Winning Combo: Draft Horses and Draft Beer

IN CONSTANT DEMAND, the eight-horse Budweiser hitches are nearly always on the road. Experienced travelers, the horses log more than 40,000 miles every year. Yet brewery officials still have to turn down more than 5,000 invitations a year.

PARADES, stock shows, rodeos, festivals, grand openings, fairs and centennials.

You name it and the Budweiser Clydesdale horses have been there. In fact, they've been practically everywhere in better than 40 years on the road.

The muffled clump-clump of 32 hoofs and the rolling rumble of the big, heavy steelclad beer wagon wheels signal the arrival of the big hitch—always a crowd pleaser regardless of location. Promoters from coast to coast say the big draft horses—averaging over 2,000 pounds each—attract crowds better than flypaper attracts flies.

Two Parade Hitches

In constant demand, the ever-traveling Budweiser Clydesdales—made up of eight-horse hitches working out of St. Louis, Missouri, and Merrimack, New Hampshire—travel 40,000 miles a year. There's also a small family of Clydesdales in residence at Busch Gardens "Old Country" family entertainment center adjacent to the company's Williamsburg, Virginia, brewery.

But up until just a few years ago, only one Budweiser hitch carried the entire beer promotion load.

There's no way the brewery can meet the tremendous demand for the Clydesdale horses each year. While the horses are on practically a never-ending schedule, better than 5,000 invitations have to be turned down

Denny Silverstein

Anheuser-Busch

each year from promoters.

Draft horses have long been identified with the beer business. And that's fitting, since, in the early days of brewing, horses supplied most of the power to get the grain to the brewery and then to later haul the finished brew out to the corner taverns.

In the days before refrigeration and pasteurization, a brewery's local marketing area was usually measured by how far its horses and wagons could travel in a day... and still get back home by nightfall. Then Prohibition came along in January of 1919.

Horses Were All Sold

Like most other breweries, Anheuser-Busch closed its stables soon after and sold its horses, since it had no work for them. During the 14 years of Prohibition, Anheuser-Busch stayed in business by making commercial yeast for brewers and bakers, as well as sarsaparilla and many non-beer products.

About a dozen years after the start of Prohibition, August Busch, Jr. was still looking forward to the day when the firm would be back in the beer business. As that day neared, he prepared a surprise for his father

---

"A short-eared horse is usually short on brains. Short ears and narrow between the eyes— stay away from 'em..."

---

by purchasing some Clydesdale horses and having them trained to pull a beer wagon. His idea was to have a shiny new brewery wagon pulled by the most magnificent horse hitch ever assembled ready to present to his father when Prohibition was repealed.

Wanting something that was entirely different and in a class by itself, Busch decided the traditional half-dozen horses pulling a beer wagon wouldn't be enough. Instead he opted for an eight-horse hitch.

Better Than Detroit Creation

On April 8, 1933, the day after the United States—enroute to full repeal of Prohibition—legalized the manufacture and sale of 3.2 beer, Busch asked his father to join him on Pestalozzi Street outside the St. Louis brewhouse for a look at his son's new car.

There wasn't any shiny new car parked on the street outside the brewery. But what was there on the street certainly stopped August Busch, Sr., in his tracks.

Instead of the expected new automobile, there stood eight magnificent Clydesdales with their gleaming leather and brass harness hitched to a shining new bright red Budweiser beer wagon!

Since that early spring day in 1933, there has always been an

---

"A horse with a shallow middle is hard to feed and makes a particularly poor traveler..."

---

eight-horse hitch of Clydesdales at Anheuser-Busch.

The beautiful Clydesdale breed has an interesting background. They originated in the River Clyde area of Scotland. Known as the "gentle giants" because of their favorable disposition, these horses were first bred by Scottish farmers during the early 18th century.

Always a Street Horse

Draft horse breeds such as the Belgian and Percheron were more popular than Clydesdales among farmers during the heyday of animal horsepower in American agriculture. But the Clydesdale was always right at home—and in much evidence—applying his snappy style to the hauling of beer, liquor, meat and other products in the growing cities and towns of America.

Farmers found the long silky feather on the Clydesdale legs objectionable. These feathered legs were often caked with mud and dust when the Clydesdale was used for farm work. Yet the feathered leg was a badge of distinction on the urban pavement where mud and dust weren't the enemy they were out

THE MUFFLED clump-clump of 32 hoofs and the rolling rumble of the big, heavy steelclad beer wagon wheels signal the arrival of the big hitch. Soon the eight Clydesdales, the Budweiser wagon, the driver, his assistant and "Bud," the frisky Dalmatian, roll into view.

on the farms of America.

The Clydesdale's springy action, coupled with the popping hocks and cleverly folded knees, made them a status symbol supreme.

The popularity of the Budweiser Clydesdale horses has grown ever since they were first unveiled back in 1933. Television, radio and magazine commercials featuring the Clydesdales and the brewery wagon have increased their popularity over the years.

The horses mean a great many things to a lot of different

ORDINARY SHOES available from standard suppliers will not fit the huge Clydesdale hoofs. Each Clydesdale shoe is hand-made from steel that measures 22 inches long, 1½ inches wide and ½ inch thick. A finished shoe weighs 4¾ pounds.

Denny Silverstein

people. The Budweiser Clydesdales are the first horses some city kids have ever personally seen "live". And some old-timers standing right next to these kids along a downtown parade route can remember when beer and freight wagons were drawn by similar horses in their hometowns.

Better than 200-million people a year now see the Budweiser Clydesdale horses in person or via television in their own living rooms.

Need Perfect Horses

To earn a place in the Budweiser hitch, or as one of the two spare horses that accompany each hitch, a Clydesdale has to have a perfect bay coloring and white markings.

Only geldings are used in the hitch. Geldings travel better than mares, and they are also less irritable than stallions.

A mild disposition is a "must" for hitch horses, since they are subjected to all kinds of abuse. This includes petting, fingers poked into the flesh or eyes, blaring horns, flashing lights, popping balloons, plus screaming children and human beings who don't get out of the way of a ton of horseflesh swinging toward them at an extended trot.

Men, Horses and Dog

A six-man crew cares for the Budweiser Clydesdales when they travel ... and that's most of the time. A driver, an assistant driver and four horsemen travel with each hitch. Besides the men and horses, a Dalmatian dog that sits on the wagon during parades completes the crew.

Each of the hitches travels in a caravan of three specially-built vans measuring 40-feet long and 8-feet wide. Two of the vans are used to transport the ten horses. A separate van is used to transport the 3½-ton brass-trimmed wagon, better than $30,000 of harness, portable stalls, feed and tack for the big horses.

Most of the feed isn't hauled along—it's purchased as they travel. That's understandable when you consider that, on the road, *each horse* consumes 60 to 65 pounds of hay and 25 to 30 pounds of oats, wheat bran and beet pulp *each day*. They also drink 10 to 12 gallons of water daily, and relish a regular snack of a pound or so of carrots.

Mighty Fancy Stabling

On the few occasions when the main hitch is at home, the Clydesdales are housed in a large red brick building next to the St. Louis brewery. Distinguished as a national landmark by the Department of Interior, this stable contains stalls for the horses, wagon storage and exercise rings.

Likewise, the Budweiser hitch in New Hampshire has a special stabling area close to the New Hampshire brewery.

A Big Clyde Breeder

To make sure the Clydesdale breed continues, Anheuser-Busch has a breeding herd located at Grant's Farm in St. Louis County, Missouri. There are usually 100 head of "gentle giants" on hand at the farm at any one time.

HITCHIN' TALK

"Wheel", "swing", "body" and "lead teams" probably don't mean much to a person who hasn't spent some time around draft horse hitches.

Yet each of these words pertains to a special team that makes up the hitch ... and each team has a special job to do.

The "wheel team" consists of the two horses closest to the wagon. Normally these are the two biggest horses in the hitch, and they control the direction and movement of the wagon. Steering the wagon with the tongue, these "wheel team" members are the only horses that can hold back when the driver wants to slow the wagon down. They are also the only horses in the hitch that can back the wagon.

The "swing team" is in the center of a six-horse hitch. This is the toughest spot in a hitch to work a green, untrained horse. A free-swinging pole that carries the pulling power from the front team to the tongue extends between the two "swing team" horses.

It is easy to swing the pole around and literally spill an untrained horse, say experienced hitch drivers. In working a new horse in this swing team, a driver tries to bump him easily a few times with the pole, until he gets the hang of turning. After a horse has worked in the swing team a few weeks, he'll have the hang of it.

The "lead team" horses are out front in a hitch. Here, you need a pair of horses that knows enough to step out quickly and really move when the driver passes the word.

In an eight-horse hitch, the extra team is known as the "body team". This team is positioned in the second position, directly ahead of the wheel team. Both the "body team" and "swing team" have a free-swinging pole between them in an eight-horse hitch.

VETERAN driver Floyd Jones put his Clydesdale unicorn hitch through its paces in the show ring. The Bangor, Wisconsin, farmer has shown Clydesdales for many years at shows around the midwest.

WITH A QUEEN at his side, Roland Ruby moves his six-horse Belgian hitch out for another daily parade at the Wisconsin State Fair. Ruby—who has enjoyed a more than half-century love affair with Belgians on his Brookfield, Wisconsin, farm—is driving a 5,600 pound wagon that is over 100 years old. Ruby says draft horses are like a disease— once you get them in your blood, you can't get them out.

Frank Lessiter

Frank Lessiter

J.C. Allen and Son

Abernathy Photo Company

EVERYONE loves a parade, including these beautifully matched grey Percherons.

"4-H HITCH" was the name for this eight horse hitch of ton-weight Clydesdales. Each year, the famous six-horse hitch of the Union Stock Yard and Transit Company of Chicago was converted into an eight-horse hitch to head the 4-H Club parades at leading fairs and the late November International Live Stock Exposition. These hitches promoted the Chicago Stockyards where more than $2 million worth of livestock was sold each day when the yards was in its heyday.

185

43 Years of Selling Draft Horses

IF ARNOLD HEXOM is known for only one thing in his life, it has to be his mastery of selling draft horses. From his teen-age days to the present, this Waverly, Iowa, auctioneer has spent a lifetime selling draft horses.

Ask him to estimate how many horses he has sold in 43 years of auctioneering and Hexom just laughs and shakes his head. He doesn't have any idea how many horses his eyes have seen bought and sold during all those years.

But he quickly recalls how he got started as a teen-ager by buying, selling and trading six to a dozen horses each year. And he's still at it, stronger than ever.

His auctioneering business has now grown to where he probably sells—or helps sell—better than 2,500 horses in a year's time. And that's a sales record he has kept up over the last ten years, as the popularity of the draft horse was rekindled.

Two Big Events Every Year

Of course, Hexom's big events each year are the semi-annual Waverly Horse Sales. Held during April and October for more than 30 years at the Waverly sale barn, the twice a year sale is the "granddaddy of all horse sales". And as Hexom recently told a visitor from western Canada, "Running this sale is the highlight of my life."

The sales generally run three days in both spring and fall. During one of those "super horse

"I'M THE ONLY auctioneer in the world that could sit up there on the block for 25 straight hours without going to the latrine," says Arnold Hexom, shown in the center.

Frank Lessiter

Dale Stierman

AT 13, ARNOLD HEXOM bought his first horse—a black Percheron colt for $49. Buying and trading, he had $465 by the time he was 15. His older brother left his money in the bank instead of "foolishly wasting it". Then the bank went "belly up".

Dale Stierman

sales", Hexom and his partner in the horse sales, auctioneer Bill Dean, have sold as many as 700 draft horses, 700 saddle horses and a whole deluge of horse-drawn machinery, horse equipment, tack and antiques.

If you're looking for a good draft horse for breeding or field work, a flashy show horse, a big horse to enter in a pulling con-

---

"I've never been to a baseball, basketball or football game, never shot pool and never bowled a line. But I've been to thousands of horse shows and sales . . . and that's enough for me."

---

test, a saddle horse, manure spreader, fancy buggy, horse-drawn grain drill, harness, saddle or whatever . . . the Waverly Horse Sale is the place to be.

The sale has become so popular that the 300 prime seats in the tent are sold out at $5 each. This is so the guys with the money to buy can find a choice place to sit. "Of course, we credit the $5 seat price toward the cost of any horse they buy," adds Hexom.

4,000 Folks Show Up

It's not unusual for 4,000 horse buyers and "lookers" to show up for part or all of the three-day sale. A recent sale attracted horse buyers from 19 states and four Canadian provinces.

While hundreds of men, women and children come to buy horses, thousands of other people come just to watch the horses and the bidders. The sale crowd includes horsemen, local farmers, horse traders, wives, kids, townsfolk, dudes, dandies, city slickers, agri-businessmen, truckers, tourists, Amishmen . . . and who knows who else.

Many Amish farmers come to buy draft horses to operate their farms. They still farm with real horsepower instead of the four-wheel high horsepower mechanized versions used by most farmers today.

Regardless of your preferred breed, weight or color, you'll find horses at every sale that make you want to reach down deep into your wallet and buy, buy, buy.

If you've now got the idea that this is no average horse auction, you're right. Thousands of dollars trade hands during the three-day sale as hundreds of draft and saddle horse fanciers buy and sell.

It seems like the horse sales start out at a fast pace . . . and then speed up. The pace is hectic and exhausting.

49 Hours of Selling

During a recent spring sale, the auction of 700 draft horses got underway at 10 a.m. on a Thursday morning. The buying and selling was continuous until the auctioneers and the crowd took their first break from selling the big horses at 6 a.m. on Friday

Dale Stierman

Dale Stierman

AMISH BOYS look over part of the big selection of horses in one of Arnold Hexom's semi-annual sales. During a recent three-day sale, Hexom sold 1,200 draft and saddle horses in 49 hours.

THE PACE of the three-day horse sale is hectic. With the exhausting speed at which the auction runs, it's good to catch a few minutes of pipe-smoking relaxation out back.

morning... some 20 hours after the first big horse was led into the sale ring the previous day!

After a four-hour break, the auctioneers were back at it at 10 a.m. "We sold draft horses that second day until 5 p.m.," recalls Hexom. "At the same time we had another sale going on the grounds where they were selling horse machinery, horse equipment and antiques."

After the draft horses and horse equipment sale was completed late that Friday afternoon, the auctioneers, sales force and crowd took the night off to rest. Then the sale of 500 saddle horses began Saturday morning at 10 a.m. These horses weren't all sold until 7:45 a.m. Sunday morning!

Hexom, the master auctioneer, directs it all. He's a non-stop guy... running at full speed nearly 24 hours a day during the big horse sale. And for a guy who left the schoolroom permanently at the age of 12, he can sure educate you about the art of selling horses.

"I Could Sell Forever"

Hexom's record for non-stop selling is 14 hours. That record was set when he was running one of the biggest twice-a-month dairy cow sales in Iowa. "We would regularly start selling at noon and sell straight through to midnight," he remembers. "Shucks, that wasn't nothing at all. My personal record is 14 straight hours of selling."

Yet Hexom figures he could do even better if he was pressed to do it. "I bet I'm the only auctioneer in the world that could sit up there on that auction block for 25 straight hours without going to the latrine. I'd bet a thousand bucks on it," he explains.

Nobody's about to take his bet with the endurance he has shown in running these non-stop horse sales.

There used to be many more draft horse sales each year at Waverly. It wasn't unusual to hold three or four sales during the winter as farmers got anxious to buy new work horses before the spring field work season rolled around.

"There were lots of draft horses in the country in those days," says Hexom. "But the sales weren't as big as we have now. We would sell 100 to 125 horses at a sale. We were having a sale every month or two when horses were really going good.

"We were really busy in the 1940s when everybody was replacing the horse with the tractor. Some of these horses went back out to farms to work the fields. But many of the horses went straight to the glue factory."

The Thrill of the Sport

While horses are the main attraction at the Waverly sale each spring and fall, Hexom keeps busy the rest of the year handling farm auctions. "Two years ago, I sold 254 auctions in a 12-month period," he proudly states. And that's keeping busy in anybody's book.

Put simply, Hexom is *addicted* to the horse business. Besides the two big horse sales, he regularly "cries" special sales or dispersals in four or five different states and Canada. He is also in demand as a master of ceremonies for horse shows and horse pulling contests around the Midwest.

Besides this all-out schedule, Hexom also finds time to raise horses of his own. He usually has 40 to 50 horses of assorted breeds, size and color on hand. When we talked with him recently, he had 30 Percherons, a few Belgians, some saddle horses, several Morgan horses, five Hackney horses (not Hackney ponies) and a team of mules roaming the home farm.

Hexom has developed a keen interest in showing his Percherons. He takes in several of the major draft horse shows each summer and fall.

"I sold my interest in the sale barn a few years ago just so I could have more time to devote to my horses," he explains. "You

Dale Stierman

might say I finally decided my job was getting in the way of my hobby!

"I always had wanted to show draft horses. But the sale barn and my farm auctioneering got so big it was a full-time operation. So I sold off my interest in the sale barn to my partner, Bill Dean."

While he has sold horses for better than 40 years, Hexom's greatest thrill in the draft horse business came in the show ring. That was when his Percherons walked off with top six-horse hitch honors at the Wisconsin State Fair a few years ago.

"I've had a lot of great moments over the years in the draft horse business," concludes Hexom. "But that win in the show ring had to be my greatest thrill."

Dale Stierman

Dale Stierman

WITH PRIME AUCTION SEATS selling for $5 each, it's more than just a place to buy a horse. Better than 4,000 folks often show up.

"YA...I'VE got a bid right here in the second row," shouts this happy ringman. The bidding goes fast and heavy as the hundreds of horses move quickly through the sale ring.

SELLING OF draft horses at a recent sale started at 10 a.m. on Thursday and continued non-stop until 6 a.m. on Friday. After a four-hour break until 10 a.m., the auctioneers sold draft horses until 5 p.m. In 27 hours of auctioneering, 700 draft horses were sold.

191

When Horses Hauled the Freight

BEFORE TRAINS and trucks, draft horses kept America rolling.

And later, even the freight that went by train was hauled to and from the station by wagons and horses.

Practically anything you can think of was hauled by horses before trains and trucks appeared on the scene.

If you looked down a busy city street in the old days, you could often count 50 or more horses hitched to a vast array of wagons loaded with all kinds of freight and merchandise from apples to zebra hides. Horses and their drivers could be seen patiently waiting their turn to load or unload at many a freight dock.

Horses were found everywhere in the old days. At the turn of the century for example, there were an estimated 40,000 horses pulling freight wagons on Detroit's streets.

A team of horses and a husky driver—plus maybe a helper or two—could deliver a tremendous amount of freight in a long day's time. Most horses were usually

FROM FIGHTING FIRES in 1910 to hauling newlyweds today, horses have kept busy. These newly-weds are being driven by Ray Bast of Richfield, Wisconsin.

Milwaukee Journal

Brown Brothers

H. Armstrong Roberts

HORSES EVERYWHERE greeted visitors to this scene of many years ago in New York City. Streetcars completed the madhouse of activity in this produce and brewing area in the city known far and wide as the "Big Apple."

H. Armstrong Roberts

FOR SHORT HAUL freight delivery in the center of big cities, the horse-drawn wagon held its own for a number of years after trucks first rolled into the picture. But eventually these freight-hauling horses lost their jobs too.

A BREWER'S market area used to be determined by how far a team of horses could haul beer in a day's time and still be back at the brewery before nightfall.

Jos. Schlitz Brewing Co.

CENTRAL CITY tavern operators relied on keg beer to get their tasty brew. Note how this wagonmaster was able to carry extra kegs between the wheels of his beer wagon.

Jos. Schlitz Brewing Co.

Milwaukee Public Museum

H. Armstrong Roberts

ICE CUTTING was an annual ritual on many a lake or pond years ago. Horses were used to clean away the snow, power the ice saws and to haul the ice to storage as in this Milwaukee, Wisconsin, scene.

PACKED IN sawdust within large ice houses, cakes of ice would last through the warm summer when ice sales were "really hot."

H. Armstrong Roberts

Library of Congress

WINTER OR summer, selling ice was a daily job that could turn into an "ice cold one" at certain times. Yet on a hot summer day, there was a certain bit of pleasure in watching a class of young school children and their pretty teacher cool off by licking a small piece of ice. Ice men often sold their product by the pound rather than by the cake.

197

entitled to a good night's rest after a hard day's work in the streets.

The draft horse more than earned its keep when it came to street work. The horse became an expert at delivering freight from store to store, milk from door to door, or beer from tavern to tavern. It often needed no driving once the route became established—a good driver could often coax his horse along from door to door with just a kind word or two. Thanks to the "horse sense" of his horse, many a milkman saved thousands of steps this way as he delivered milk in the wee hours of the morning.

Delivering all kinds of freight, mail, groceries and milk was often hard work for both the horses and the driver. But it made the taste of a bag of oats and a good cold drink all the better when quitting time rolled around.

HAULING WATER out of a creekbed, these eight horses and two drivers had an easy, quick way to cool off on a hot summer day.

PACK TRAINS were a frequent sight moving across the Great Plains as they moved large amounts of freight to customers in the new part of our country. This ten horse team was moving west with the two drivers riding on the side of the wagon instead of up top.

STRINGING WIRE for telephone and telegraph lines was handled by this crew. Dressed up in what looks like their Sunday best, they were not wearing their working clothes.

SHORT OF WATER, this farmer dipped from a river bottom to fill his barrels so he could water his stock.

H. Armstrong Roberts

H. Armstrong Roberts

J.C. Allen and Son

Milwaukee Public Museum

POLE LOGGING is the specialty of Cliff Nix and his son, Lawrence. With a Belgian mare named Suzie and an old red chain saw, the two men thin trees that they saw up and sell as poles or corral posts. Cliff has worked 35 years in the woods. "I never pay much attention to time," he says. "Don't even have a time piece. Never thought my time was worth much."

John White

IT'S A BIG LOAD containing 31,480 board feet of lumber weighing 113 tons. The 63 logs, stretching 21 feet high and 20 feet wide, was hauled one mile in 1892 by Wisconsin's Ann River Logging Company. The sled runners were 5 inches thick, 11 inches high, 9 feet long and were spaced 9 feet apart. Just the sled and chain alone weighed five tons. The biggest load of logs ever hauled by two horses was a 59.2 ton load that contained 36,055 board feet of lumber.

State Historical Society of Wisconsin

The Woodsman, a Saw and His Horse

IN A DAY'S TIME, a typical pair of sawyers could cut their way through 100 pine logs. The logs would be skidded into a holding area by a single horse to await shipment out of the woods

HERE AND THERE, the draft horse still hangs on in the woods. But it's becoming even less here and less there, as the push is for bigger and bigger powered machines in the woods.

"Clear cutting" of timber probably had as much to do with the demise of the horse in the woods as anything.

When timber barons were doing "selective" cutting of isolated

State Historical Society of Wisconcin

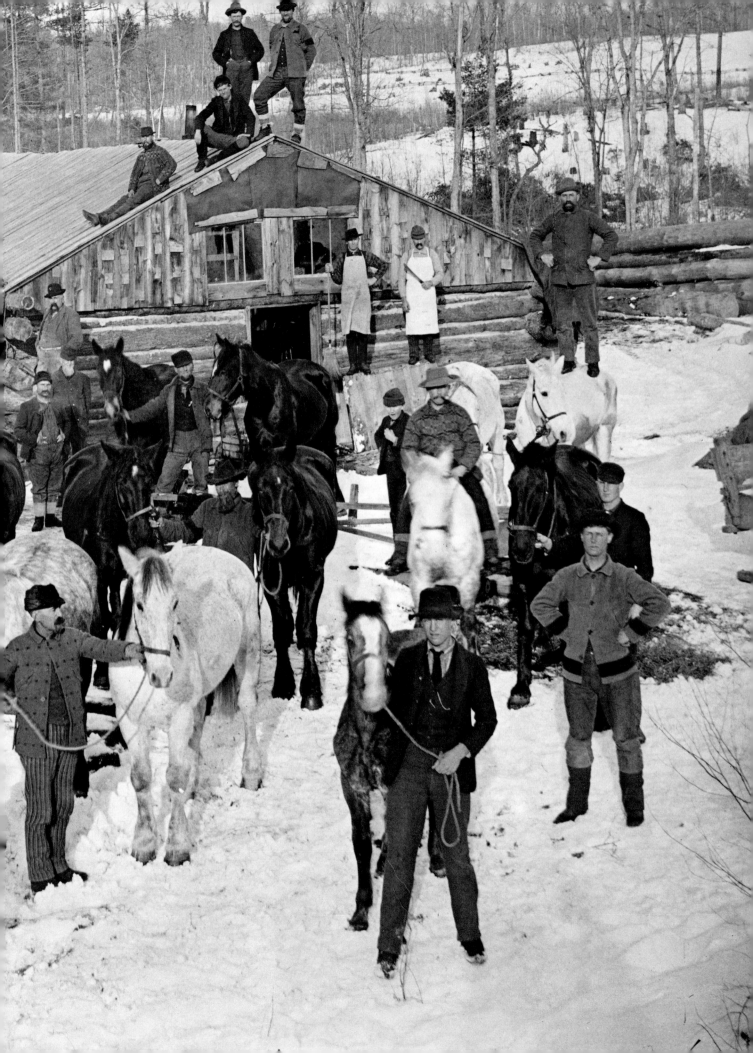

trees in a forest, the horse had some real advantages in dragging out these trimmed logs through limited space. The horse could maneuver better in these close quarters than large machinery.

An added benefit was that horses didn't freeze up in 25 below zero weather, and their parts didn't shatter in the harsh cold—operators of some mechanical logging monsters will testify to the latter problem even today.

While there has been a big shift to mechanization of woods work, you'll still find a horse here and there, such as the ones shown here working in the woods.

ON THE PREVIOUS PAGE, 30 men struck up all kinds of interesting poses when the photographer showed up to take this northern Wisconsin logging camp winter-time portrait.

Milwaukee Public Museum

WORKING DEEP in the woods, Cliff Nix says timber cutting is the only trade he's ever known. Having spent 35 years working in the woods, Nix used to work in snow three or more feet deep. But he's given that up in recent years since it is too hard to cut trees in those kinds of conditions.

John White

Where the Horse Is Still King

MACKINAC ISLAND is one of the few places in the world where the draft horse still rules the roost.

To get around the island, you have no choice but to "get a horse"... unless you're willing to walk or pedal a bike. You can't just switch on the ignition of a car or a truck, because *motor-powered vehicles have been banned from the island since 1928.*

As a result, one of the things that immediately strikes a visitor to Mackinac Island is the absence of truck and car noise. There are no traffic jams, no screeching brakes, no racing motors and no honking horns to be heard anywhere.

And the air? It's as pure and free of automobile exhaust fumes and factory smoke as you'll find anywhere.

All you can hear is the staccato beat of the big horses' hooves hitting the pavement as they pull carriage wagons and horse-drawn taxis laden with tourists. It's a comforting sound in a nostalgic setting.

Officially, residents of the island (located halfway between Michigan's upper and lower peninsulas) admit that there are a few "gas-burners" on the island. But each of those are the necessary type that fall in the emergency vehicle category... an ambulance, police car (most policemen ride bicycles), and fire trucks. There is also a snow plow, and electric and telephone company trucks, but they are used only in the winter months.

So the "hay-burners" definitely have the "gas-buggies" out horse-powered when it comes to numbers on Mackinac Island.

Yet most of the 450 draft

COLLEGE KIDS make up most of the crew of 50 drivers who handle the summer carriage tours.

Frank Lessiter

HARNESSING AND unharnessing 200 horses a day keeps everyone on their toes. Some 285 horses work on the island during the summer.

Frank Lessiter

and saddle horses that are ridden for pleasure or pull carriages, taxis and freight wagons take up residence on the four square mile island only during the summer.

A few other means of transportation exist, but the horse is king. Bicycles are seen everywhere during the summer season, when 5,000 to 15,000 tourists a day arrive on the boats that make the 20 to 30-minute trip from Mackinaw City or St. Ignace docks. And there are a few riding horses on the island. But that's about it.

The 600 folks who stay on the island the year-around have been known to start up a snowmobile or ski down to the post office to pick up their mail on peaceful winter days. And a few horse-drawn sleighs can also be seen on the streets during the winter

"No traffic jams, no screeching brakes, no racing motors, no honking horns . . . just draft horses and bikes . . ."

months. Plus, there's an airstrip in the center of the island that has proven valuable numerous times during winter emergencies, when the island is isolated from the rest of civilization by the frozen waters of Lake Huron.

Two Barns, 285 Draft Horses

Two big barns, a blacksmith shop, a downtown taxi horse barn, a few acres of exercise lot and 285 draft horses make up the island operation of Mackinac Island Carriage Tours. This firm keeps horses on the island from early May to mid-October to pull the tour carriages and taxis that serve the hundreds of thousands of visitors during the summer

Frank Lessiter

Frank Lessiter

IT TAKES 1500 tons of hay a year to feed the 285 horses. The horses spend their summers working on the island, then go north to Michigan's Upper Peninsula to rest during the winter.

CARRIAGE tour licenses have been held by some of the same families since the late 1800s. Incorporating in 1948, the remaining 30 carriage owners have worked together since that time.

207

Frank Lessiter

WITHIN ONE minute, the barn crew has the morning team unhitched and the afternoon team hitched up.

Frank Lessiter

77 YEAR OLD Sam Schneider drove a carriage for 12 years while the little red mare on the right spent 25 years on the wagon. "She and old Sam really liked to work," says barn superintendent Bob Gippespie. "Try to give Sam a day off from driving and he would fight you."

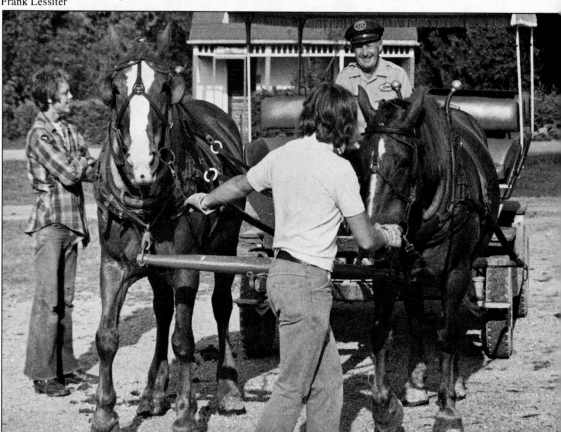

Frank Lessiter

A HORSE costs $750 and harness runs $400 each for 285 animals, says Jim Chambers. Plus, there's the buggy, barns, 70 men, boarding house, feed and vet bills.

island's essential taxi service.)

By 7 a.m. each morning, most of the 50 drivers will have bicycled the one mile up to the barns from the downtown company boarding house to start bringing in 50 teams from the exercise lot. These horses are grained and harnessed.

By 8 a.m., the first tour carriages will be heading out to

"It's hard work when you hitch and unhitch 100 teams in 90 minutes every day of the week..."

---

meet the boats that start arriving at 8:30 a.m.

After the 50 teams and carriages hit the pavement, the 20-man barn crew brings in another 50 teams from the exercise lot, grains them and harnesses them up. (These men also unharness the morning teams in mid-afternoon, haul out four loads of manure a day and load the exercise lot racks with two and one-half tons of hay during their 12-hour workday.)

Artists of the Quick Hitch

About 1 p.m., the barn crew starts switching teams on the carriages. Each carriage driver will pull into the horse barn yard for a stop midway in his 90-minute tour of the island. As the 10 to 12 tourists seated in the tour wagon watch in astonishment, the barn crew has the morning team unhitched and the afternoon team hitched up in less than a minute's time. Then the tour driver hollers "Giddap!", the team lurches ahead and the tour is again on its way around the island.

Members of the barn crew will

tourist season that visit the island.

"When we are running full-tilt in the middle of the summer, we will have 100 teams out on the tour carriages on any given day," says Jim Chambers, president of the firm. "Each carriage needs two teams a day, so we are harnessing at least 200 horses every day during the summer."

Besides the carriage wagons, the firm runs a dozen horse-drawn taxis. These taxis, equipped with two-way radio, operate 24 hours a day the year-around. (From late October to early May, they hire one of the freight companies to handle the

209

RUBBER CUSHION shoe on the right is compared with a normal horseshoe. The carriage horses wear out $24,000 of shoes each year.

TWO BLACKSMITHS keep the horses pounding the pavement. Working on 25 or more horses in a day's time, these "blackies" can whip through a foot trimming and reshoeing in 20 minutes to get the horse quickly back out on a carriage tour.

tell you it's hard work when you unhitch and hitch 100 teams in less than 90 minutes every day of the week.

On-the-Spot Repairs

Whether it's a broken piece of harness or a new shoe that's needed, some crew member can handle it. The operation has two blacksmiths to keep all the horses pounding the pavement. These two "farriers" often work on 25 or more horses in a day's time. A harnessed horse may be led into the blacksmith shop with a foot problem, then find himself back on a wagon within 20 minutes following a foot trimming and a new shoe or two.

Most horses go through at least two pairs of shoes during the four months they work on the island. Some go through six or more pairs of shoes—particularly if a horse happens to have the habit of dragging his foot over the hard paved roads.

Yet these aren't ordinary horse shoes that the Mackinac Island horses wear. They wear a special rubber cushioned shoe with a steel insert developed years ago by island horsemen. Designed for the carriage horses

---

"They go through better than $24,000 worth of horseshoes every year on the island..."

---

that continually work the island's paved roads, it offers better traction and more cushioning than a typical rubber horseshoe.

The annual horseshoe bill isn't cheap, even though the company pours its own rubber shoes. Other horse owners pay $16 per pair for the shoes. With the carriage tour operation probably going through close to 3,000 horseshoes in a season, the bill would come to better than $24,000 of horseshoes each year at that rate.

Horses Go North for Winter

While they don't head for Florida or California to enjoy the winter sunshine, the horses only spend summers on the island. Each fall, the horses are loaded on the tourist boats to St. Ignace where they are trucked to winter headquarters near Pickford in Michigan's Upper Peninsula.

At the farm, they run and graze lush pasture until cold weather and snows arrive in late November. Then they are housed in barns during the winter months.

"In late summer, right after Labor Day, we cut back to 30 carriages and start shipping 16 to 24 horses a day to the Upper Peninsula," says Jim Chambers.

70 MEN keep the horse operation going every day. Besides the 50 carriage tour drivers, a 20-man barn crew handles the daily chores. Mainly seasoned draft horse hands, many of these barn crew members return year after year.

the same island house where their father and grandfather were born. Four generations of the family have lived on the island since the original family members arrived in 1862.

Half the Drivers Come Back

About half of the 50 carriage wagon drivers return each summer. Several retired men usually drive every year, plus a number of the college-age drivers come back for two or three years of driving. A school teacher or two may also fill their summer break

"The last of the horses are loaded on the boats and we shut down the island operation completely by late October."

Used to Be Rough Going

Some of the carriage licenses have been held by the same island families since the late 1800s. While the carriage owners formed a loosely-knit corporation in 1924, it didn't answer their wide-open competition

> "Horses wear a special rubber cushioned shoe with a steel insert that offers better traction on pavement..."

problems. Trouble continued for many years.

"In the old days, you would see 55 buggies lined up on the street as the first boats arrived in the morning," says Bill Chambers, a St. Paul, Minnesota, veterinarian who was raised on the island and who now joins the horse crew each summer. "It was cut-throat competition in those days. There were many fights and it was always rough-going."

Following World War II, high insurance rates and the continued rough competition forced the remaining 30 carriage owners to agree to incorporate. This was done in 1948. Most of the original 30 carriage horse operators still own stock in the corporation.

After being away from the island horses for 16 years, Bill Chambers started returning to the island for the summer draft horse season after his father died. He handles carriage dispatching duties and does the veterinary work on the horses.

Both Bill and his brother, Jim, were born in the same room in

---

NEED A LIFT, MISTER?

To get a ride on Mackinac Island, you just hail a horse-drawn taxi just like you would a cab in any big city. Simply wave, whistle, telephone or try anything else to get the driver's attention.

The taxi is an open-air wagon with a top to keep out the sun, rain and snow. During the winter months, taxis are enclosed to help passengers keep warm. Taxi service is available 24 hours a day every day of the year.

By contrast, the 50 open-air carriage tour wagons that carry tourists around the island handle up to a dozen passengers each. These roofed tour wagons—known as the "surrey with the fringe on the top"—travel established island routes and don't carry taxi passengers.

There are also rent-by-the-hour horses and carriages available with either a liveried driver or as a true-life "U-drive it experience."

The glamorous Grand Hotel—with its 261 rooms, 350 employees and famed 880-feet long flag-decorated porch—has a number of rather elite carriages to take guests to and from the boat dock. One of these carriage is valued at $25,000. The hotel keeps 16 horses to pull these carriages.

by driving the tour horses.

Barn superintendent Bob Gillespie says they train 20 to 25 new drivers each spring. They hitch an experienced 14 or 15-year-old team that knows all the tour ropes to a "school bus" carriage and take a dozen new "student drivers" out at a time to learn their lessons.

An experienced driver handles the team and "shows and tells" the new men about the tour as they drive the route. Later, the drivers learn to harness and clean their teams.

"Actually most of the horses are trained so well they could go out and show the new driver the whole route," laughs Gillespie. Besides driving, each driver harnesses his morning team, unharnesses his afternoon team and keeps the horses clean.

Who make the best drivers? "We like to hire mostly college kids," says Gillespie. "They fit in best here during the summer. But we want a kid with some size to him so he can throw on the harness.

"Within two or three weeks, we sort out the timid ones who are scared of the horses. We suggest they get another job both for their sake and ours since there is no sense having them spend the summer working with horses if they don't like it."

The 18-man barn crew is made up mainly of seasoned draft horse hands. A number of these men return year after year.

Feed Comes by Boat

It's mighty expensive to feed horses on the island, since all of the feed must be shipped in by boat. Cost of hay during our visit with Gillespie last summer was $3 per bale . . . with boat charges from St. Ignace running $15 per ton. A 96-pound bag of oats cost $9.

And plenty of hay is used! The horses go through 600 tons of hay on the island and another 900 tons of hay a year at their winter headquarters.

You'll see Belgian, Clydesdale and Percheron breeding along with a sprinkling of some of the lighter horse breeds in the island horses. A lightweight draft horse weighing 1,300 to 1,400 pounds is preferred for carriage wagon work. Heavier horses can't take the severe pounding from the hundreds of miles on the island roads each summer.

While the corporation breeds a few of its own horses, most are purchased. And a good carriage

NO GAS STATIONS

What if you decided to bring a truck or tractor to the island? Could you do it? Would they let you?

Well . . . you had better have a darned good reason. And even then you can't be sure of getting permission.

To drive any motor vehicle on the island, you need a permit from the City Council. And if the road you are going to drive on is bordered at any point by state-owned property, you also need a permit from the Island Park Commission.

When a motor vehicle moves (such as a construction company truck or tractor), it must be preceded by a team of horses and freight wagon attached to the motor vehicle. You can have the motor running and the gears engaged to actually make the truck move. But it's got to look like the horses are doing all the work.

And there's usually a policeman on a bike in front of the horses. That's for protection since island folks and horses aren't used to the "loud noise" of an internal combustion engine.

In addition, you will have specific hours and days spelled out for use of the motor vehicle. Chances are the hours will be late at night . . . so the islanders can pretend your truck or tractor really isn't there.

CARRIAGE WAGONS cost $4,000 each. The big investment in this operation has to be earned back during the tourist season.

SNOW TIRES work best on carriages. Regular tires go flat as the tires don't roll fast enough to kick pebbles out of treads.

AN IDEAL carriage horse weighs 1,300 to 1,400 pounds. The best carriage horses they ever had were half Percheron and half Arabian.

Frank Lessiter

Frank Lessiter

Frank Lessiter

213

horse team for island use will cost $1,500 or more these days.

Half Percheron, Half Arabian

Gillespie has fond memories of three horses he considered to be ideal carriage horses. "We bought some lightweight horses that were half Percheron and half Arabian from Michigan State University years ago when they sold their draft horses," he recalls. "We got three of them and they weighed around 1,350 pounds each.

"That Percheron-Arabian breeding combination really made a great carriage horse. But who can afford to breed a horse like that today with the high prices being paid for both Percherons and Arabians?"

EVERYTHING ON the island moves by horse or bike, including 98,000 pieces of luggage each summer. College student "dock hops" quickly master the knack of balancing nine suitcases on a bike.

Frank Lessiter

Frank Lessiter

Drays, Horses and Freight

THINK OF FREIGHT and the idea of thousands of big horsepower diesel and gas-powered trucks speeding down the nation's highways often comes to mind.

That's the way freight is hauled and handled in most places. In most places, that is. But on Mackinac Island, other means of freight hauling have to be found, since trucks and cars are banned from the island.

So, Mackinac Island residents rely on four-legged horsepowered drays (freight wagons) to deliver everything they need to run their businesses and homes. It allows you a close look back at the way a great deal of freight was delivered in years gone by, especially in the downtown areas of many big cities.

Three family-owned freight lines operate on the island hauling fuel oil, horse feed, lumber, gravel, groceries and anything else you can name from the boats. It all goes to eight hotels, five rooming houses, 18 restaurants, 34 shops, four liveries, four bike rental firms, four churches, a sometimes operating college and many private homes.

Cowell Enterprises, headed by Reuben Cowell, is one of the island's three freight lines. Cowell's father settled on the island in 1948, originally working draft horses at the Grand Hotel before starting his own freight line.

Cowell, his wife and six children now live and haul freight on the island the year around. Cowell's teen-age son is

RUBEN COWELL keeps five teams and wagons busy all summer long hauling freight. Many summer days run from 6 a.m. to 10 p.m.

Frank Lessiter

ALL FREIGHT moves to Mackinac Island by ferry boat, except for a few items that may arrive in a small airplane during the winter months. Each of the nearly 100 boats that arrive with tourists on a hot summer day also carries freight of one type or another. When the snow gets deep in the winter, the freight lines deliver goods on sleighs.

Frank Lessiter

Frank Lessiter

LINES TIED TIGHT to the wagon brake keep the team from moving while freight is loaded or unloaded. Within 15 to 20 minutes of its arrival on the island by boat, freight is on its way to a final destination.

the third generation to work in the family freight business, having driven horses since he was old enough to hold the lines.

Summer Pay Days

"It's a pretty tough life," explains Cowell. "When you live here, you make it all in the summer or not at all. You work hard in the summer—often from

RUBEN Cowell's family has hauled freight on the island since 1948. His teen-age son, the third generation to work in the family business, has driven horses for years.

"OIL TANKERS" are hitched right behind the freight wagons. During its rounds, the freight crew will drop the 500-gallon tanker containing fuel oil off at a house or store. The emptied tanker will be picked up a day or two later.

Frank Lessiter

Frank Lessiter

HEY HAUL THE FREIGHT IN and Cowell's crew also hauls the rubbish away for a number of the Island's 18 restaurants, 34 shops and 8 hotels. Rubbish ends up at a landfill site in the center of the island —also the final resting place for the island's never-ending supply of horse manure.

Frank Lessiter

TIGHT CORNERS often require some careful maneuvering on the part of the teamsters. But it's just part of the job of delivering freight everywhere.

THREE FAMILY- owned operations compete for the island's freight business. They haul anything you can think of . . . and more. One of Ruben Cowell's teams once towed a smashed-up airplane down to the boats.

Frank Lessiter

Frank Lessiter

Frank Lessiter

Frank Lessiter

6 a.m. until 10 p.m.

"But you have all winter to rest. While we haul some freight in the winter, it's still pretty much like taking a long vacation when you compare it to the hectic hours and hard work in the summer."

Freight Every 30 Minutes

Cowell keeps five teams and wagons busy during the hectic summer months hauling freight. Freight arrives practically every 30 minutes from morning to night, as nearly 100 passenger-carrying boats a day float in from Mackinaw City or St. Ignace during mid-summer. There is freight on board nearly every boat.

Switch to Sleighs

From mid-December until mid-April, freight arrives by airplane. When the snow gets deep, Cowell uses sleighs to deliver freight. He also runs the taxi service for Mackinac Island Carriage Tours during the winter months when they take their horses to Pickford for winter boarding.

While the Grand Hotel has its own battery of draft horses for hauling, Cowell takes over for them in early spring and late fall when their horses are also back on the mainland.

Cowell prides himself on being able to haul anything anywhere on the island. "We once towed a smashed up airplane from the center of the island down to the boats after it had

---

"There is nothing on the island I don't haul. You name it— I've hauled it . . ."

---

crashed," he recalled. He has also hauled the President's luggage during visits to the island.

When freight is moving every which way on a busy summer day, Cowell still takes time to grain his horses during the noon hour. "Hot humid weather is mighty hard on a horse," he says. "If you don't play out a team, they can do a lot of work for you. We've found a team of work horses is never stuck until the teamster is."

CAT-NAPPING on top of some laundry bundles, a teen-ager grabs a few minutes rest. Summer days can be busy, but the long winter days when the tourists are gone and life on the island grinds to a halt can be like a five-month vacation.

Frank Lessiter

COLLEGE studies will start again in a few weeks for this student as his summer of work on the island draws to a close. But regardless of what profession he enters, this summer of working with horses will never be forgotten.

FANNING ITS tail, a horse may be helping spread the aroma of the famous island fudge. Shopowners use fans to blow the cooking fudge smell out into the street to lure tourists inside to buy.

Frank Lessiter

WHILE MANY summer residents go south in the winter, the horses move north into Michigan's Upper Peninsula for the winter.

PULLEYS AND ROPE allowed a team to pull a loaded hay fork to the top of the barn for dumping and mowing away. Many hay rope teams were driven by young boys and girls, but few got to ride in style as did this young Iowa gal.

J.C. Allen and Son

J.C. Allen and Son

A YOUNG COWBOY'S work was never done. Even "Hoot Gibson" types had to sometimes lay down their chaps, holsters and guns instead of riding off to a gunfight.

When Farm Kids Horsed-Around

BAREFOOT BUT HAPPY, straw-hatted Jimmy was pleased when Dad handed him the lines and told him to head Jake and Sam to the barn all by himself.

FARM KIDS in the old days were almost as inseparable from the farm's horses as today's kids are from their pet dogs and cats.

Many horses and farm kids enjoyed a mutual respect built on love, understanding and the fact that they often had to work darned hard together.

Many a farm boy found him-

J.C. Allen and Son

FAMILY portrait on the Paul Mitchell farm near Battle Ground, Indiana, many years back resulted in a couple of smiles, a frown or two and one outright case of bawling. We can't read the Percheron's face for sure, but it looks like he's hoping this will be the photgrapher's last shot.

J.C. Allen and Son

IF HE'S OLD enough to hang on, he's old enough to drive, believed some old-time teamsters. Hauling hay shocks—pulled by running a rope around the hay— fell on the shoulders of this young farm boy.

J.C. Allen and Son

Dale Stierman

J.C. Allen and Son

AMISH farmers still rely on real horsepower to do their field work. Handling the lines during grain bundle pitching time was a job for this young but barefoot Amish boy.

FEEDING OATS to his favorite colt, this young lad seemed to be enjoying it all. Looking on without being the least bit jealous was the boy's dog.

225

WHAT MARE could resist trotting up to the fence for some hay out of the hands of a cute young gal.

COLT CLUBS helped many young farm boys learn the basics of raising good livestock.

J.C. Allen and Son

J.C. Allen and Son

J.C. Allen and Son

SHOWING OFF their favorite colts for the photographer, these two cousins seem to be enjoying it all.

226

HEADING HOME after a long day's work in the fields, a young boy is entitled to a bareback ride and the chance to rest his tired and sore feet.

J.C. Allen and Son

J.C. Allen and Son

J.C. Allen and Son

SCARED OF these two big brutes? Never! After all, this young farm gal grew up with these two Percherons.

BUT THE BOY'S not so sure. He'll feel better when Dad helps him down and asks if he would like a drink.

self handling the lines of a team at a young age. This freed an older brother, father or hired man for tougher farm jobs.

This author fondly remembers the many times he guided the team down the alfalfa windrows while dad and the hired man forked hay away from the hay loader. Or driving the wagon in wheat and oat fields while the others pitched and stacked bundles. Or...shucks, my list of horse driving jobs could go on forever.

Dale Stierman

HEADING HOME, this Amish boy has had an enjoyable day visiting neighborhood friends.

A.M. Wettach

H. Armstrong Roberts

THE HAY'S UP so let's head back to the barn and start over as soon as the men folk are ready.

PA'S PROUD of the work these three turn out. Junior already handles the team as well as Dad.

SHORT PANTS and no shirts on these young boys signal another hot summer day on this farm. The boys will soon be climbing down and cooling off with a sip of lemonade in the shade while Dad and the two big greys head for the field.

AS THE BOYS get older, their interests change... to things like taking their favorite gal for a Sunday afternoon drive through the beautiful farm country.

J.C. Allen and Son

H. Armstrong Roberts

Medicine, Harness and Horseshoes

A TEAMSTER can't do it all alone.

Besides himself and the horses, he needs a good backup team of skilled people to keep the horses in top-notch condition for work, play or show.

Most of these men call in experts when it comes to keeping the horses healthy, the harness in good repair and the horses properly shod. Yet, some horsemen still do much of this work themselves when they can.

"Call Dr. Lundvall"

When treasured draft horses get real sick or pop a tendon in a leg, "Doc" Lundvall will often get a telephone call.

Dr. Richard Lundvall is the head of the equine clinic at Iowa State University. He and his students treat 5,000 to 6,000 draft, race and pleasure horses each year in this Ames, Iowa, clinic. It's not unusual for a teamster to haul a problem horse 300 to 400 miles so Lundvall can take a look at it.

When Lundvall graduated from veterinary school back in 1944, the bulk of the clinic's horse business was draft horses. Even then, draft horses were being phased out in favor of tractors for farm work.

"So for a number of years, we treated very few draft horses here in the clinic," recalls Lundvall. "But eventually the draft horse started making a comeback... and today we treat a lot

SHOEING HORSES has always been hard work. Blacksmithing hasn't changed much over the years and still takes plenty of strength.

J.C. Allen and Son

"DOC" LUNDVALL is likely to get a call anytime a teamster finds a valuable horse with a "bum leg."

Frank Lessiter

WITH A REAL love for harness-making, school teacher James Stampone, Jr., would like to build his new business to where he could give up his Maine school desk for good. Stampone taught himself harness-making in order to keep his Belgian team rolling. He charges $400 to $600 for work harness and up to $2,500 for show harness. The final price depends on how much brass you want and just how fancy you want to get.

Judith Buck Sisto

of the big draft horses."

Feet Problem Prone

Lundvall says the most common medical concerns with draft horses are feet and legs. While veterinarians can surgically correct some foot and hock problems, a great deal of the work must be done through corrective shoeing.

"Despite its size, the draft horse is not very well footed or hocked," says Lundvall. "The bulk of the draft horses today are Belgians... and they are notoriously poor-footed.

"While the Clydesdale has good feet, they are not as good as the Percheron. In fact, the Percheron has better feet than any other draft horse breed."

School or Harness?

As for harnesses, every little town used to have its own harness shop in the old days. But that was when this country had lots more work horses.

Today, harness shops are few and far between. One that is far from most places is run at Freeport, Maine, by a full-time school teacher and part-time harness maker that goes by the name of James Stampone, Jr.

Admitting that his real love is harness making, Stampone hopes to build his business to where he can forget school teaching altogether and spend all of his time making harnesses.

"I'm basically a self-taught harness maker," he says. "I have

"I learned harness making by patching harness for my own team..."

a team of Belgians and learned about harness making from patching up their harnesses and by reading books."

A working harness can cost $400 to over $600 today, depending on how fancy you want to get and how much brass you want on it. And for a show harness, the price can go as high as $2,500!

Stampone makes both show hitch harness and pulling harness and there are distinct differences between the two types.

"A pulling harness has a lot heavier traces," he says. "Pullers use about a 3-in. trace, where you may only use a 1½-in. or 1¾-in. trace on a show hitch harness.

"A show hitch harness uses a lot of patented leather and fancy stitching. It's also prettied up more and that costs money."

As harness leather, Stampone buys what the trade calls "harness backs." These are large pieces of cow leather with the belly cut off. Coming from fairly old steers, the leather is at least ¼-in. thick in the center of the back.

"You cut your traces right out of the middle of the back," says Stampone. "Then you cut the other parts of the harness out of

MENDING HARNESS was a bigger source of income for most harness shops than the making of new harness. Farmers often sought miracle jobs on harness that probably belonged in the trash heap.

J.C. Allen and Son

H. Armstrong Roberts

the leather down toward the belly. This lower leather is very stretchy and soft, so you can use it on parts of the harness that don't take any stress."

A Farrier Afoot

Proper horse shoeing takes real skill. Jim Rupple really means it when he tells you he has been shoeing horses ever since he graduated from high school.

"I graduated from high school on Friday and showed up at horseshoeing school the following Monday," says the Medina, Wisconsin, farrier.

"Shoeing horses is something I always wanted to do. When I was in high school, I worked with the Roland Ruby Belgians at Brookfield, Wisconsin. Now I shoe horses for them."

After learning the ins and outs of the forge, hammer and shoe, Rupple spent two years shoeing and caring for the Budweiser horses in St. Louis, Missouri.

Returning to Wisconsin in 1975, he started his own horseshoeing business. Yet Rupple still continued to shoe many of the Budweiser Clydesdales.

This often required journeying around the country to keep the traveling horses in shoes. Like two trips to New Orleans within seven weeks.

"I flew down the first time, since the shoes only had to be refitted," says Rupple. "But I drove my truck down the second time, since I needed the forge and a ton of stuff to put on a brand new set of shoes. That was quite a trip with my old truck—2,400 miles roundtrip to be exact."

With eight hitch horses and two spares to shoe, Rupple usual-

---

"I graduated from high school on Friday and started horseshoeing on Monday..."

---

ly needed three days to complete the job. Since he usually stopped by St. Louis to shoe other spare horses on the way, the whole job usually took about a week's time.

Big Feet or Small Feet?

The way a farrier shoes a horse depends on how the horse is used. "For draft horses going into the show ring, you try to get the biggest foot you can," explains Rupple.

"A hoof will spread out more by August fair time if it is trimmed and shoed in February, May and July. A draft horse looks its best in the showring when it has a really big foot."

But it's a different story for pulling horses. Here, Rupple says, you want the smallest foot you can get. "You want a foot with all of the weight being pulled from one small area. A small-footed pulling horse is not so likely to stumble."

The amount of corrective shoeing Rupple does depends on the particular horse. "While there are not many horse legs I can't straighten, you must be careful since you run the risk of hurting a tendon or hock if you go too far."

Regardless of what you might think, Rupple doesn't get kicked much. "I am usually close enough to the horse that I don't get hurt much even if he does kick—it's more like a push," he says. "Plus, most horses can't

THE YEAR WAS 1929 as a white Percheron mare stood waiting for new shoes outside this picturesque Connecticut blacksmith shop. Besides shoeing horses, this blacksmith spent a considerable amount of his time building wagons.

J.C. Allen and Son

ROPES ARE IN PLACE to hobble this horse so his feet can be treated.

HOT SPARKS FLY as a shoe is heated so the proper fit can be made.

Frank Lessiter

Frank Lessiter   Frank Lessiter

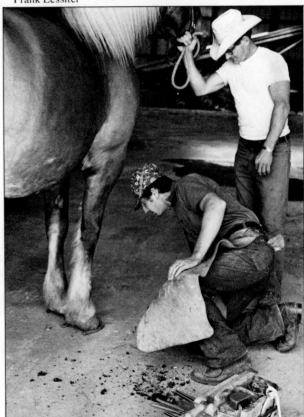

WITH SHOW HORSES, you want to get a big foot. But you want a small foot on pulling horses.

JIM RUPPLE shoes horses mainly in eastern Wisconsin. Yet he used to travel anywhere in the country where the Budweiser hitch was on tour to keep the big Clydesdales in new shoes.

Frank Lessiter

kick you once their foot is picked up off the ground. They need the other three for balance. But there are exceptions."

Needed: 800 Horses

While a number of farriers earn good money, horseshoeing is not a get-rich-quick scheme. It also takes a pretty sizable investment to attend horseshoeing school and to buy needed equipment.

Bob Bechdolt charges $1,200 for tuition, board and room for his Girard, Kansas, school. The shoeing course includes 60 lectures and 60 laboratory periods. Talking with Bechdolt, you hear some pretty colorful stories—all a result of 21 years of shoeing horses for the Army.

His school's dozen students

EXPERIENCED FARRIERS often turn to teaching, like Robert Reaume.

Bonnie Pollard, Michigan Farmer

never see a live horse until halfway through their 12-week session. Yet, a dozen students will shoe over 800 horses before they graduate.

Standing sweaty and black by his forge, Robert Reaume, another horseshoe school owner, works with the intensity of a man who cares about his craft. Wiry and lean, this owner of the Wolverine Farrier School at Howell, Michigan, clangs away with his 2½-pound hammer, shaping a reluctant piece of red-hot steel on his anvil.

Clad in the traditional leather chaps, Reaume opened his school when he got tired of the constant travel that went with shoeing horses.

"It's like being a doctor on call," he says. "If somebody's prize horse cracks a foot a few days before a big show, it doesn't matter that you have other things to do. They expect you—in fact, *demand* that you go there and take care of that animal."

Like Bechdolt, Reaume han-

AN OLD-TIMER once told Tony Castagnasso there are two ways for a blacksmith to go to hell—by hitting cold iron and by not charging enough. When Tony shoes a horse, he starts from scratch with a 22-inch piece of steel. With a hot forge, he pounds out four shoes in three hours.

dles a dozen students at a time in his 12-week course. He paid his way through Michigan State University by shoeing horses in his spare time.

He furthered his blacksmithing knowledge by taking a college horseshoeing course from Jack MacAllen, a legend in the draft horse business.

Reaume is a man of many talents and temperaments. One moment he may be banging away furiously at a poor-fitting shoe. The next moment he is discussing the artist Goya.

Soon after, he pauses to expertly propel a stream of tobacco juice into the forge where it immediately hisses into steam.

He's highly talented in all three areas.

YOUNG TOM weighed 2,900 pounds when Darwin Starks bought him. With Tom still growing bigger every day, Starks figured the horse had a good chance to beat the 3,200 pound world record weight for a horse as listed in the Guinness Book of Records. The big horse stood 20 hands high, equal to 6 feet, 8 inches in height.

Frank Lessiter

His Goal: The World's Heaviest Horse

SINCE HE WAS a teen-ager, 6 ft., 3 in. Darwin Starks has had a dream of someday getting his name in the record books as the owner of the biggest horse in the world. And he's working on it.

The Sun Prairie, Wisconsin, farmer and horse trader first got interested in draft horses when he was 14 years old. This interest has continued to grow ever since.

A Lifelong Desire

Searching for a big team to buy a few years ago, Starks heard about two huge horses from a fellow horseman. "But when I called the owner in the Blue Ridge Mountains of West Virginia, he told me the team was not for sale," recalls Starks.

Still interested and curious, Starks and a friend headed to West Virginia to see the team anyway. Starks asked the owner if he would take $10,500 for the two big Belgian geldings named Tom and Dick. The owner, a logger, said he would have to think about it.

They agreed to meet several weeks later at an Ohio horse sale. And Starks spent much of the next few weeks busily scraping together $10,500 in $100 bills to take with him to the Ohio horse

DICK AND TOM weighed a total of 5,700 pounds when Starks bought them for $10,500. He later sold Dick into Indiana for $10,000.

Wally E. Schulz

auction to buy the horses.

"When the time arrived to head for the Ohio auction, I was scared to pack that much dough on me," explained Starks. "So I wore two pairs of pants that day and kept the cash pocketed in the inside pair."

When Starks arrived late at the sale, he learned the logger had already left for home, since a bad ice storm was moving into the mountains. Undaunted and determined despite the icy roads, Starks and a friend drove their big truck to the logger's home.

When they arrived at the mountain house, the logger was doctoring another horse's leg. Putting the horse's leg in a sling, the logger was heartbroken over the possible loss of this horse.

Starks has always felt since that this incident is what may

Frank Lessiter

> "If we could put 301 pounds of gain on the horse, the world's record was ours..."

have swayed the logger to part with the big team. The logger knew the same kind of unfortunate injury could happen to either of his big horses in the woods... and he would end up with nothing.

So after Starks tossed those 105 separate $100 bills on the kitchen table, the logger decided to sell. They counted out the bills, totalling $10,500, they shook hands, loaded the horses and Starks was on his way back home with the big team.

Got Investment Back

Starks weighed the horses when he got home. Dick weighed 2,800 lbs. and stood 19 hands tall. (A hand equals 4 in.) Tom weighed in at 2,900 lbs. and stood 20 hands high (6 ft., 8 in.)

After doing some exhibiting of the Belgian team around the Midwest, Starks eventually sold Dick to an Indiana farmer for $10,000. So he practically had his total investment back from the sale of just half of the big

logging team he had bought.

Shortly after he sold the one horse, Starks set a personal goal. "I decided to put some more weight on Tom and beat the 3,200-lb. world record weight for a horse as listed in the Guinness Book of Records."

Starks figured he had a good chance to set a new record with Tom. After all, the young horse weighed 2,900 lbs. at that time

THE WORLD'S HEAVIEST HORSE

The Guinness Book of World Records shows the heaviest horse ever recorded weighed 3,200 pounds.

Named "Brooklyn Supreme," the purebred Belgian stallion stood 19.2 hands (6 feet, 6 inches) tall. He died in 1948 at the age of 20 at Callender, Iowa.

The tallest horse ever recorded was "Firpon." He stood 21.1 hands (7 feet, 1 inch) and weighed 2,976 pounds. He died on an Argentina ranch in 1972.

A BIG, BIG HORSE was the only way to describe Tom. His huge size is evident when you consider that Darwin Starks stands 6 feet, 3 inches.

and was still growing rapidly.

"I'd put 550-lbs. of gain on another horse the previous winter, taking him from 1,960 lbs. up to 2,510 lbs.," Starks explained. "So I figured I could easily add enough weight to this big framed horse to get my name in the world's record book."

With that, Starks started feeding the big horse toward a record weight. And Tom proved to be a big eater—he ate 60 lbs. of shelled corn and 50 lbs. of hay each day.

No Record This Year

Things looked good for setting a new record when this writer first talked with Starks. But when I talked with Starks six months later, he was not as optimistic about setting a new world's record with Tom.

"The horse got a bad wire cut on one of his legs and it's really slowed him down," said Starks then. "We have no idea how the

DARWIN STARKS SHOWS just how big Tom was from his neck down to his hoofs. Starks paid $100 for a pair of 34 and 36 inch horse collars. The seller didn't believe there were horses anywhere big enough to wear these collars. Since the biggest shoe available wasn't large enough, Starks always had to add more steel.

wire cut happened. But it has really set him back. He lost quite a bit of weight before we got the bad infection cleared up.

"While Tom is now back up around 2,900 lbs., I don't believe he is going to beat the present world's record this winter."

At that time, Starks figured the horse would weigh around 3,000 lbs. by April 1. But hot

---

"It cost us $946 to feed the horse for just seven months..."

---

summer weather would keep him from gaining the 200-plus pounds of needed weight before the following fall.

Still, Starks wasn't discouraged—the horse was still young, and Starks figured he could still feed him to a world's record the next year. Plus, following in his father's horse trading footsteps, Starks figured he might also sell him. "But I wouldn't take a penny less than $10,000 for him," he told me at that time.

We wish we could provide a happy ending to this story and tell you Starks fed his horse to a new record weight and that he got his name in the Guinness record book. But we can't.

Later that same summer, tragedy struck. Late one Sunday morning Starks wandered out to the horse barn and found Tom lying in the stall with a broken leg. The leg could not be properly set and the horse had to be shot. Starks buried him right on the farm.

What the future holds for Starks and his goal of getting his name in the record book is uncertain at this time.

But one thing is for sure, Darwin Starks still has his eye set on getting his name into the Guinness Book of Records for owning the biggest horse in the world.

And with the determination he's shown to date, we wouldn't bet against his doing it one of these days.

Eastward Ho the Wagons: American History in Reverse

TRAVELING across the country in a covered wagon is something many people have heard their ancestors tell about. But for one group of draft horse owners, it's more than just history—they *experienced* that piece of history.

Frank Lessiter

WHILE PIONEERS had it tough years ago when they headed west to the promised land, their modern-day counterparts insisted on more comfort when they headed back east to celebrate our nation's 200th birthday. Yet the 1976 trail still proved plenty tough.

That's because these modern-day teamsters made the trip themselves. They packed bags, bedrolls, horse feed, wagon grease, food and much more to take part in a unique 1976 coast-to-coast wagon train pilgrimage that helped celebrate our nation's 200th birthday.

It was a never-to-be-matched adventure that moved eastward along five separate routes, from as far west as California and the Canada-Washington border to Valley Forge in Pennsylvania. Some western Canadian draft horse owners even joined up for part of the trip.

The Bicentennial wagon train was a replay of history—but in reverse. While our ancestors journeyed west on the long, sometimes dangerous 19th century trip in covered wagons in search of the promised land, these modern-day pioneers journeyed east in 1976.

Wagons From All 50 States

Each of the 50 official canvas-covered Conestoga wagons—one from each state—was historically authentic, except for modern-day steel axles, roller bearings and hard-rubber tipped tires mounted on the wooden spindle-wheeled wagons.

The re-creation was a thrill for more than just the participants. The wagon train was a beautiful sight as spectators watched the lead outriders come into view

Aitkin-Kynett Co.

Buck Miller, The Milwaukee Journal

ROLLING INTO TOWNS for the night, the wagons drew thousands of visitors who wanted to see how their ancestors had traveled. Below, John Purciker from Fort Atkinson, Wisconsin, takes a break from wagon train driving duties.

Frank Lessiter

over a countryside hill. Next came as many as 48 canvas-covered Conestoga and other assorted types of unofficial wagons, being pulled by draft horses, mules, ponies, Quarter Horses and oxen.

These were followed by 50 to 100 outriders—in the saddle for a day or more of hard riding—enjoying their participation in an event they will recall for many generations to come.

It was a scene right out of the past. The only reminder that it was taking place in the middle of the 20th century instead of 100

EVEN THE GEESE FLOCKED to see the Bicentennial wagon train as it made its way across America toward Valley Forge. The wagon train was an instant hit with everyone wherever it went.

Al Herrmann, Jr., The Pittsburgh Press

years earlier was the brightly-lit police car out front, the portable radios blaring out country music, or a glimpse of some other modern convenience that showed up every now and then as the wagons and riders flashed by.

The pace was fast—they had to stay on the move to make the required 12 to 40 miles a day. It seemed the wagon train was quickly past you before your eyes had a chance to take it all in. Many spectators would dash to their cars and hurry down

---

"Fifty Conestoga wagons—one from each state—moved coast to coast toward Valley Forge..."

---

back roads to get ahead of the train for a second look.

The wagon trains originated in Washington, California, Texas, North Dakota and Maine. "Branch trails" from 43 other states also funneled into these main trail routes.

13 Months on the Trail

The northwest segment of the train set out from the Canadian border on June 6, 1975, with 500 riders and 30 wagons. A month later at the Oregon border, a number of Washington riders and wagons turned back as fresh riders and wagons from Oregon joined up.

This scene was repeated throughout the pilgrimage as 18 states were crossed on the trail from Washington to Valley Forge. Yet, some diehards like Pat Doran made the whole 3,500 mile trip.

Doran, a riding stable operator from Blaine, Washington, footed most of his own expenses since he was not driving one of the

Frank Lessiter

Frank Lessiter

SLOW MOVING VEHICLE signs were one of the few changes made on the wagons from days gone-by. This was the way many of our ancestors moved west.

EXTRA COMFORT was enjoyed by some folks along on the wagon train for the whole trip or just a few days of fun. Rubber-coated wheels and seats from an old auto made the bumps more bearable.

EVEN WITH JAMMED-TOGETHER WAGONS, the parade often took 25 to 30 minutes to pass. There were the official wagons, 15 to 30 unofficial wagons and 100 or more outriders in the saddle for a day or two of hard riding along the trail.

Frank Lessiter

official Bicentennial wagons.

He had to rely on a pickup truck as a brake to help his 14-mule team ease down a particularly steep section of the Cascade Mountains. "I'm not sure I could have been a real pioneer in the old days," he confessed. "It must have been rough."

Ed Porritt, an artist from Green River, Wyoming, also made the long trip east. "My great-great-grandmother walked from Iowa City to Salt Lake City pulling a handcart with four children in it," he says. "I thought a lot about that and got out of the wagon and walked every time I could. I figure I walked better than 1,500 miles on the trip.

No Indian Attacks

Each evening, the wagon drivers drove their wagons into a big wide circle around the campfire—just like their ancestors did. Only this time around, there was no real need to do this for protection against Indian attacks.

This gave people in the host towns and cities along the way a chance to meet the wagon drivers, look over the wagons, watch the feeding and grooming of the horses and be entertained by a special wagon train band of singers and actors.

While visitors would linger on into the late hours of night, the wagon drivers and their families tried to turn in early. They had to be out of the sack by 5:30 a.m., eat breakfast, have their horses fed, pack, harness up and be on the road by 8 a.m. While they traveled as few as 12 miles and as many as 40 miles a day, they averaged around 20 miles per day on secondary roads and along historic trails, where pos-

MOM CARRIED the water and feed buckets to the county fairgrounds barn where the horses rested for the night while dad unharnessed the team. Traveling as a team, many farm couples dressed up in early-day clothes to make their journey eastward even more authentic.

AFTER NINE HOURS and 29 miles on the road, the teams are unhitched and unharnessed for a night of rest. But soon morning will arrive and the team and teamsters will again be on the trail that brings them closer to Valley Forge each day.

ALL THREE GENERATIONS will remember the enjoyable days spent with the Bicentennial wagon train for years to come. Traveling with the wagon train was an unforgettable opportunity to take part in an All-American event that may never be repeated.

sible, before bedding down for another night in another small town.

Three Sleeping Bags

Young Ron Cole spent many a night sleeping on the ground around the campfire on the way to Valley Forge. "I used three sleeping bags when it was cold," says the 16-year-old ranch hand from the McCrossan Boys Ranch at Sioux Falls, South Dakota. "I would put one bag underneath me, crawl into one and toss the third bag over the top.

"The early morning hours really got to you. The frost

"The constant traveling was harder on the drivers and their families than the horses . . ."

---

would be all over you. When you woke up, you got up in a hurry!"

Traveling was harder on the people than the horses. Many folks felt the horses would have feet and leg problems, but this didn't occur, even though horseshoes lasted an average of only a few days with the continued pounding of the road.

"We had bones that hurt where we never knew we had bones," says Art Hood, a rancher from Orange Springs, Florida. "But it was worth it."

As the wagon trains wound their way through the small towns of America, drivers, passengers and outriders often found themselves being treated as instant celebrities.

"We All Felt Important"

G. A. March, was an unemployed fence builder and factory guard when he signed on as a member of the southern wagon train. "It was more exciting than a circus," he says. "Every night people came out to talk to you, asked for your autograph and made you think you were *somebody* when you knew you really weren't."

Al Mavis, an Illinois man on the wagon train, says it appeared that to many spectators the trip represented the *real* Bicentennial. "You could hardly go to bed at night—people just wanted to stay, sit around the campfire and sing," he says.

The Mississippi Horse Council tapped Monroe Magee to be wagon master through the deep South. "I wouldn't have missed this opportunity to travel with the wagons," he says. "We had a big time and public response was tremendous. We had 3,000 to 5,000 people at nearly every campsite."

There were many benefits to making the trip. Keith Kreykes headed up as many as 48 wagons carrying 150 people during his term as wagon master for part of the trip. "You forgot what day it was and forgot the time pressures," says his wife, Gail. "You were able to live each day to its fullest."

The End of the Line

The wagon train pilgrimage terminated on July 4, 1976, at Valley Forge Park in Pennsylvania on the 200th anniversary of our nation.

It was estimated that over 150,000 people—from babies to folks as old as 85 years of age—rode in the wagon trains at various stages of the trip. Besides the folks riding in the wagons, thousands of outriders joined the train for one or more days along the way. Additional teamsters joined up with horses and wagons for a day, a week or a month or more at various points along the trail.

Some people on the trail found things haven't changed too much since their ancestors went west in covered wagons 100 years ago.

"Back in the pioneer days, a man could travel across the country and find himself welcome anywhere he went," says Jay Eubank of Bryan, Texas. "People would offer him a place to stay, something to eat and treat him like a friend. On the trek to Valley Forge, we found these things hadn't changed a bit."

All kinds of folks—a minister, school teacher, veterinarian, grandmother, neurologist, teenager, to name a few—were along for at least part of the wagon train ride.

Spending more than four months on the way to Valley Forge was Bryce Johnson of Scranton, North Dakota. A rancher all his life, Johnson left the family livestock operation in his wife's hands while he was gone.

"We got the lambing out of the way before I left on April 2 for the second leg of this journey," he says. "Then one of my sons came back to the ranch to handle the calving. We hired field work done in the spring and summer, just so I could take part in this once-in-a-lifetime experience."

Johnson's wife didn't mind staying home and tending the ranch. "She likes to get away from the ranch, but she always wants to be home by sundown," he explains.

Honeymooned in Wagon

College students Fred Shivers and Debora Johnson were married on February 28 in front of the Oklahoma state wagon. They had planned to marry in June, but moved up the date when

---

"Some teamsters made the entire 3,500-mile trip . . ."

---

they saw this never-again opportunity to join the wagon train.

The couple sold two cars, riding horses, a team of mules, an antique saddle collection, plus leather items they had made at college to finance their trip.

Using this cash, they bought a covered wagon and a team of horses called "Smokie Joe" and "Kokomo." Then they lit out along the trail for Valley Forge—enjoying a four-month honeymoon along the way.

The farrier on the Mississippi wagon train was a Louisiana veterinarian named Earl LeMoine. "I figured if you can't go on

(Continued on page 254)

Frank Lessiter

TRUSTING SOULS, most wagon drivers simply left their harness draped over the wagon tongue each night. Later in the evening, many drivers would crawl into sleeping bags near the harness and wagons to spend a peaceful night under the stars.

MEAL TIME each evening meant an ample supply of hay and grain was always dished out for the horses. This feeding wagon was set up with a portable stall. Pulled ahead of the wagon train each day with a pickup truck, it was set up and loaded with feed by the time the hungry horses rolled into the campground each night.

Frank Lessiter

Aitkin-Kynett Co.

Some Things Just Never

Everything changes in 200 years, right?

Not true.

Participants in the Bicentennial wagon train found some things—some good and some bad—exactly as they were with wagon trains 200 years ago.

• Take horse manure. It smells now exactly as it did 200 years ago. And it is still true that if you walk around a wagon train after dark you are surely going to step in some of it.

• A blazing campfire on a chilly night still attracts people—they still back up to it and say how good it feels.

• Night sounds are still pretty much the same—the peepers, the rustle of tethered horses and the occasional bray of a mule. (It must be a fact of a mule's life that when it opens its mouth it has no idea whatsoever what kind of sound is going to come out. It is a sound so strange and strained that you have to wonder if the good Lord was having an off-day when he put in the mules vocal cords.)

• Is that a coyote howling? No, it's a lonesome dog. Well, it sounded like a coyote there for a minute.

• The ground is just as hard to sleep on now as it was 200 years ago. The dew moves over all things and creeps into your bedroll during the dark hours, just like it always has. Therefore, you sleep like a cat with the mange. Everytime a horse moves or a car drives past, you wake up.

• Night starts gathering up its darkest things and gets ready to move out for the coming day shortly after 4:30 a.m. By 5 a.m. 10,000 robins are singing in the hills and crows are cawing out in

WAGON TRAINS STARTED from California, Washington, Texas, North Dakota and Maine. Branch trains from 43 other states also funneled into these main trail routes.

STARTING AT THE western Washington and Canadian border, the northwest segment of the wagon train took 13 months to make the trip back to Valley Forge. Some 18 states were crossed on the 3,500 mile trip.

Aitkin-Kynett Co.

Seem to Change

the misty shadows. Soon people begin to stir here and there among the wagons, carrying water buckets or bales of hay. Where the heck are Ward Bond and Chuck Conners?

• A man comes by leading a team of horses and someone asks him if they drank. You want the fellow to say, "Yes, I led a horse to water and I made him drink," But instead he says: "A little." So much for the old familiar historical sayings.

• It's a cold, wet morning and people walk around with their shoulders humped up against the chill. Most of the covered wagon occupants have a pickup camper or camper trailer along for creature comforts. They come out of these, tend to their horses and then head over to the old school bus that has been converted into a "Chuck wagon" for breakfast.

• After breakfast, trainmaster Keith Kreykes comes around on his horse. He tells everybody the train will leave in 10 minutes.

People harness their horses and hook them up to the wagons. And the outriders ride their prancing horses in and out among the wagons.

Then Kreykes goes up to the head of the wagon train, and instead of saying, "Wagons, ho!", or "Eastward, Ho!", or something like that, he just nods to the driver of the lead wagon. The procession follows the flashing lights of a police squad car out onto the road, and another day on the trail begins.

Left behind are the wagon tracks in the morning dew, the gray ashes of the campfire, and the horse manure.

Some things never change.—*By Bill Stokes, Milwaukee Journal.*

something like this when you want to, then what good is working?" he explained.

Many teamsters joined up at the last minute. Andy Erickson of Badger, Minnesota, was one of them. He didn't have any horses or even a wagon when he suggested to his wife that they join the train.

They bought some horses and a wagon and headed east. "We had never been east before and figured this was an unmatched chance to get there," adds Erickson.

Selling the lease on the ranch he had operated for eight years, Buck Buckingham of Sisters,

---

"The pace was fast to make 12 to 40 miles per day . . ."

Oregon, used the money to build a wagon. This became the traveling home on wheels for his wife and four children for the next six months.

Good Job Had to Wait

Merle Johnson of Snohomish, Washington, turned down a $36,000 per year job in Indonesia to make the wagon trip to Valley Forge. "That job will still be there when this trip is over," he rationalized. "I just couldn't resist signing up for this ride."

Others planned for the trip well in advance. Max Preuss of Ixonia, Wisconsin, spent a full year getting his wagon in shape to spend just three days with the train through part of the Badger state.

"I figured this would be the last time I'd ever have a chance

ACHING BACKS WERE EASED on many of the Bicentennial wagons with car seats, wooden chair backs, seat pads and plush sports car bucket seats. These modern-day conveniences made the trip east more bearable for many teamsters. As one wagon driver put it, "We had bones that hurt where we never knew we had bones."

KANSAS CITY HERE I COME was the slogan for 34 Indiana Future Farmer of America members who organized their own memorable wagon train experience. These boys from 13 Indiana chapters each spent three or four days riding, walking and driving a wagon 500 miles from Indiana to Kansas City.

Lyle Orwig

to get in on it," laughs Preuss. "I won't be here 100 years from now when they might do this all over again."

He came equipped for the trip with a lot of draft horse experience. A retired steam locomotive engineer, Preuss farms 88 acres with the help of seven Belgians.

A working cowgirl from Antelope, Texas, also made the trip. Mrs. Hazel Bowen, a 68-year-old grandmother has punched cattle all her life. "I rope, brand and do everything any cowhand can do," she says. "So the wagon train ride seemed like an appropriate vacation to me."

Tom O'Keefe of Sonoita, Arizona, sold his bar, bought a century-old wagon and joined

---
"One-just-married couple spent a four-month honeymoon on the wagon train trip . . ."
---

the train with his wife. "This wagon that brought a family of seven from Mississippi to Arizona in 1875 has taught us much about ourselves," he says. "I've learned to live with wagon grease and my own self-reliance."

Saw How Ancestors Did It

Finally . . . many of the sightseers along the way watched the wagon trains roll by with amazement and recollection of stories told by their own ancestors.

"I wanted to see how my grandmother had traveled many years earlier," says Sarah Uersteeg, a spectator along the trail from Romulus, Michigan. "I remember her telling how bad the roads from Pennsylvania to Indi-

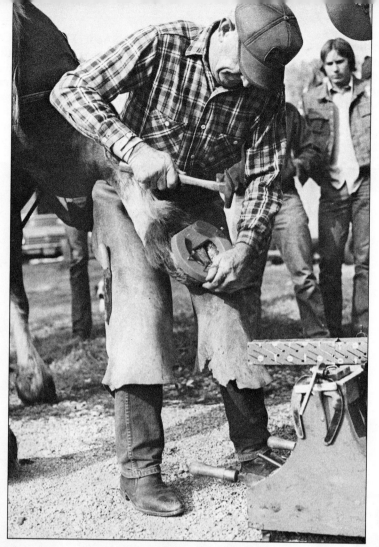

ana were—especially in the rain.

"She told me how they milked the cows in the morning, put the milk in barrels in the back of the wagon and would find butter when they stopped for the evening."

Those homesteaders who journeyed west by wagon train in the

"Horseshoes usually lasted an average of only three days..."

19th century never forgot the experience. Neither will the thousands of wagon drivers, wagon passengers, outriders and spectators alike who were part of the Bicentennial wagon train east in 1976. For most, it was truly a once-in-a-lifetime experience.

TEMPORARY SHOEING was done as needed along the way until the students could arrange for a farrier to fit new shoes. During the three week trip, the wagon was always a big attraction in cities and towns.

Lyle Orwig

Lyle Orwig